Geometry of Molecules

Chemistry-Biology Interface Series

Charles C. Price, Editor
Department of Chemistry
University of Pennsylvania

Editorial Board: Aubrey W. Naylor, Duke University; Robert W. Burris, University of Wisconsin; L. Carroll King, Northwestern University; Leonard K. Nash, Harvard University

Geometry of Molecules

Charles C. Price
Professor of Chemistry
University of Pennsylvania

McGraw-Hill Book Company

New York	Kuala Lumpur	Panama
St. Louis	London	Rio de Janeiro
San Francisco	Mexico	Singapore
Düsseldorf	Montreal	Sydney
Johannesburg	New Delhi	Toronto

Geometry of Molecules

Library of Congress Catalog Card Number 77-159314

1234567890   BA   7987654321

This book was set in Times New Roman by Textbook Services,
Inc., printed on permanent paper by George Banta Company, Inc.,
and bound by George Banta Company. The designer was Marsha
Cohen; the drawings were done by B. Handelman Associates,
Inc. The editors were Jeremy Robinson, Maureen Mc Mahon,
and Janet Wagner. John F. Harte supervised production.

Contents

The Chemistry-Biology Interface Series

Several years ago, a few dozen biologists, chemists, physicists, and other scientists spent several days on the campus of the University of Washington under the joint sponsorship of the Commission on Undergraduate Education in Biology, the Advisory Council on College Chemistry, and the Commission on College Physics. The purpose was to study ways to improve teaching in areas of mutual concern to two or more of the disciplines involved. The group considering the area between chemistry and biology agreed that a series of paperback books, prepared for elementary college level students in either biology or chemistry could serve a useful purpose toward this end.

Prepared by authorities in their fields, these books could, for the chemists, indicate the biologically significant reactions useful to illustrate chemical principals and, for the biologist, summarize up-to-date information on molecular phenomena of significance to a modern understanding of biological systems.

To implement this proposal, CUEBS and AC_3 appointed an editorial committee of:

Professor Robert H. Burris, Department of Biochemistry,
    University of Wisconsin
Professor L. Carroll King, Department of Chemistry,
    Northwestern University
Professor Leonard K. Nash, Department of Chemistry,
    Harvard University
Professor Aubrey W. Naylor, Department of Botany,
    Duke University
Professor Charles C. Price, Department of Chemistry,
    University of Pennsylvania

to organize the undertaking.

As of this writing, the following volumes have been published:

O. T. Benfey, "Introduction to Organic Reaction Mechanisms"
Roderick K. Clayton, "Light and Living Matter," Vols. I and II
Charles C. Price, "Geometry of Molecules"

The following volumes are planned:

Myron Bender, "Catalysis"
Melvin Calvin, "Chemical Evolution"
Paul M. Doty, "Macromolecules"
David E. Greene, "Surfaces, Films, and Membranes"

It is our hope that the material in these volumes will prove of sufficient interest to teachers and students in elementary college chemistry and biology courses that much of it will ultimately be incorporated in regular textbooks.

Charles C. Price
Philadelphia, Pennsylvania

Preface

It is the aim of this book to introduce college students in chemistry or biology to many of the most basic factors which influence the remarkable and important properties of materials composed of natural or synthetic polymers. Many of the most important molecules in living systems, such as proteins, nucleic acids, starch, and cellulose, are such "giant molecules." Many of the most important products of chemical industry, such as nylon, rubber, orlon, poly(vinyl chloride), polystyrene, polyethylene, dacron, poly(methyl methacrylate) (Lucite or Plexiglas), silicones ("bouncing putty"), and Teflon are also "giant molecules."

The long-chain polymer molecules are thus of great significance. Their utility and function, varying widely from the action of protein in muscle, to the genetic code reproduction of deoxyribonucleic acid (DNA), to the properties of rubbers, fibers, and films, can be related to certain features of the structure of the units of these molecules.

One of the most important characteristics of these molecules is their geometry. We will therefore examine such information as the length of bonds between atoms and the angles between such bonds. It is also important to know the minimum distances between atoms which are *not* chemically bonded to each other.

In addition to molecular geometry, it is also important to consider the forces holding atoms together in molecules. These forces include not only direct bonding but also attractions and repulsions between atoms within the same molecule and between atoms belonging to different molecules.

A third most significant factor is the dynamics of assemblies of atoms and of molecules. How does the energy of thermal motion influence bond distances and bond angles within a molecule? What effect does thermal agitation have on the weak forces between atoms or molecules not joined by chemical bonds?

After exploring these questions for small molecules in a simple, semiquantitative way in the early chapters, it will be our purpose to see how these concepts can be used to build some understanding of the behavior and properties of natural and synthetic polymers.

It is assumed that the reader is familiar with the basic concepts of chemistry and physics, at the level of high school courses in these subjects.

Charles C. Price

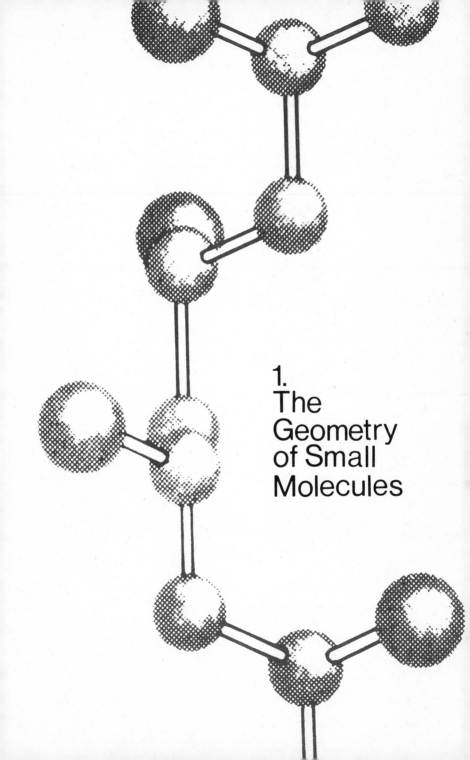

1.
The
Geometry
of Small
Molecules

INTRODUCTION

The most important molecules in living systems are the bio-polymers–carbohydrates, proteins, nucleic acids–whose properties and functions are crucial to the living systems and whose behavior is in many ways unique. These molecules share the structural feature of being chain-like molecules, constructed of repeating units of similar (or identical) structure. The properties of these biopolymers are to a large degree conditioned by their detailed geometrical shape. For example, the proteins which serve as enzymes, the remarkable catalysts so unique and vital to living processes, may lose their catalytic activity merely by a change in the shape of the molecule. The crucial function of DNA in transmitting genetic information is critically related to the ability of these long-chain molecules to form intertwined twinned helix conformations.

It is the purpose of this monograph to discuss the aspects of structure, especially the structural details of the building units of such long-chain molecules, which affect the geometry of these molecules. As we shall see, the shapes and properties of these molecules are determined in large part by intra- and intermolecular forces so that these factors also are an integral part of the discussion which follows.

As a basis for understanding the properties of giant molecules and biopolymers, it is necessary to understand such basic features of small molecules as the geometry of directed covalent bonds, including bond distances, bond angles, and bond strengths; the dynamic features of molecules, including their mobility and the stretching, bending, and rotation which can occur at covalent bonds; and the interactions which occur between atoms and molecules *not* joined by covalent bonds, including electrostatic and electrokinetic interactions.

The distinguishing feature of the molecules we shall discuss is that they are bonded together into discrete and unique structures by bonds which are not readily disrupted under normal conditions. We may contrast sodium chloride, which exists as an ionic crystal lattice, and propane, the major constituent of LPG (bottled gas). In the former crystalline substance, positively charged sodium ions and negatively charged chloride ions are attracted to each other and bonded into an orderly crystal lattice or held together in the liquid molten salt by the general, nondirectional coulomb attraction of positive charge for negative. This bonding is governed by Coulomb's law; i.e., the attractive force F is proportional to the product of the two charges ϵ_1 and ϵ_2, divided by the distance between them r (and the dielectric constant D of any medium between them)

$$F = \frac{\epsilon_1 \epsilon_2}{rD}$$

All the ions in a crystal of any size are attracted by others of opposite charge and repelled by those of like charge in accordance with this law. In other words, the potential energy is decreased and bonding occurs, compared to the ions separated by a large distance r, if the net attractive forces are greater than the repulsive forces. In sodium chloride there are no discrete molecules, and no directed bonds join just two atoms to each other.

In propane, which has the molecular formula C_3H_8, the situation is quite different. In the gas phase, propane can be shown to exist as discrete entities C_3H_8. For example, 44 g of propane occupies the molar volume, 22.4 liters at standard temperature and pressure. In dilute solutions in benzene, the molecular weight determined by the freezing-point method is 44 g mole^{-1}. In both the liquid and the crystal, the discrete units are C_3H_8.

For any covalent molecule, like propane, one of the first questions to be answered is: What is its structure? By this we mean primarily: In what sequence are the atoms attached to each other? This question can then become more and more refined and sophisticated to include bond distances separating atoms, bond angles, preferred rotation around bonds, etc. The determination of structure has always been a major part of organic chemistry, and the classical methods used are studied in organic chemistry courses, but several features of the structure of covalent compounds important to our understanding of biopolymers will be discussed here.

BOND STRENGTH

One such feature is the *strength* of the bonds holding covalent molecules together. It is not possible to account for this bonding by the simple electrostatic attraction of charged atoms, as for sodium chloride. For example, in liquid or crystalline hydrogen, the atoms are joined in pairs by a strong covalent bond, but attraction to other molecules involves only weak forces.

It is possible to measure the strength of bonds by measuring the energy necessary to dissociate molecules.

$$H_2 \longrightarrow 2H \qquad 104 \text{ kcal/mole}^{-1}$$
$$CH_4 \longrightarrow CH_3 + H \qquad 103 \text{ kcal/mole}^{-1}$$

For diatomic molecules, the bond dissociation energies are summarized in Table 1.1. The very high values for N_2 and CO correspond to the presence of triple bonds which must be broken to dissociate these molecules.

Table 1.1 Bond Dissociation Energies

Molecule	Kcal mole⁻¹	Molecule	Kcal mole⁻¹
H_2	104	HF	135
F_2	37	HCl	103
Cl_2	59	HBr	87
Br_2	46	HLi	58
N_2	226	NO	150
O_2	119	CO	256

For more complex molecules, it has been found that bond-dissociation energies for the same bond in different molecules, while not identical, do not vary widely, so that it has been useful to establish *average* thermal bond energies for the common covalent bonds, as summarized in Table 1.2. Note that these bond energies are very large compared to the thermal kinetic energy of atoms at room temperature of less than 1 kcal mole⁻¹. This accounts for the fact that molecules built with these bonds maintain their identity and do not dissociate under normal conditions.

Table 1.2 Average Thermal Bond Energies

Bond	Kcal mole⁻¹	Bond	Kcal mole⁻¹
O—H	110	C≡C	195
C—H	99	C—O	85
C—F	~110	C=O	178
C—Cl	78	C—N	62
C—Br	67	C=N	121
C—C	82	C≡N	191
C=C	146		

BOND LENGTHS

Another parameter of the covalent bond which has been determined extensively and accurately is the bond length, i.e., the internuclear distance. The most important classical methods have been x-ray and electron diffraction studies. More recently very accurate information on molecular geometry has also come from microwave absorption spectra. Values for some typical bond lengths are summarized in Table 1.3. The unit commonly used is the angstrom (Å), which corresponds to 10^{-8} cm.

Table 1.3 Bond Lengths

Bond	Å	Bond	Å
F_2	1.42	C—C	1.54
Cl_2	1.99	C—H	1.09
Br_2	2.28	C=C	1.34
HF	0.92	C≡C	1.20
HCl	1.27	C—O	1.51
HBr	1.41	C=O	1.21
H_2	0.74	C—F	1.49
N_2	1.09	C—Cl	1.76
O_2	1.21	C—N	1.52
O—H	0.96	N—H	0.93

Note that double bonds are both stronger (Table 1.2) and shorter (Table 1.3) than the corresponding single bonds, and triple bonds are even shorter and stronger.

A useful approximation to bond lengths in wide varieties of covalent molecules can be made on the basis of assuming constant covalent radii associated with atoms. These radii are somewhat shorter for double (or especially triple) bonds than for single bonds, and some of the more useful values are summarized in Table 1.4. It should be kept clearly in mind that covalent-bond lengths (Table 1.3) are experimentally measurable parameters of a molecule while covalent-bond radii are merely convenient (and in a sense arbitrary) values assigned to atoms which give approximately correct bond distances when the sum of any two radii is taken. Thus the sum of the radii for H (0.30 Å) and Cl (0.99 Å), that is, 1.29 Å, is close to the experimentally observed value of 1.28 Å for the length of the HCl bond; and the sum of the covalent radii for C (0.77 Å) and N (0.70 Å), that is, 1.47 Å, is close to the observed value of 1.52 Å

Table 1.4 Covalent-bond Radii and van der Waals Radii

Element	Single bond, Å	Double bond, Å	Van der Waals, Å
H	0.30	...	1.2
C*	0.77	0.67	†
N‡	0.70	0.60	1.5
O	0.66	0.55	1.4
F	0.64	...	1.35
Si	1.17		
P	1.10	1.00	1.9
S	1.04	0.94	1.85
Cl	0.99	...	1.8
Br	1.14	...	1.95

* Triple-bond radius 0.60 Å.
† Van der Waals radius for CH_3, 2.0 Å.
‡ Triple-bond radius, 0.55 Å.

The van der Waals radii can be estimated from information on bond distances combined with data giving molecular volumes. The volume of 1 mole of liquid hydrogen or methane can be used to estimate the volume occupied by each molecule, i.e., the volume from which other molecules are excluded. This corresponds to estimating the closest approach of one molecule to another. For example, 1 mole (6.02×10^{23} molecules) of liquid methane occupies about 32 ml, or one molecule occupies 53×10^{-24} ml. Since the volume V of a sphere of radius r is $4\pi r^3/3$, the radius is $\sqrt[3]{3 \times V/4\pi} = 2.3 \times 10^{-8}$ cm, or 2.3 Å. The value for the radius of methane calculated in this way is thus in reasonable agreement with the accepted value of about 2.0 Å for the van der Waals radius of the methyl group.

Two examples of how the covalent and van der Waals radii can be used to give an accurate estimate of the geometry of simple molecules are illustrated in Figs. 1.1 and 1.2. In addition

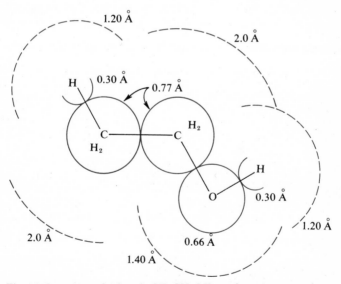

Fig. 1.1 *Geometry of ethanol,* CH_3CH_2OH, *and pertinent covalent and van der Waals radii.*

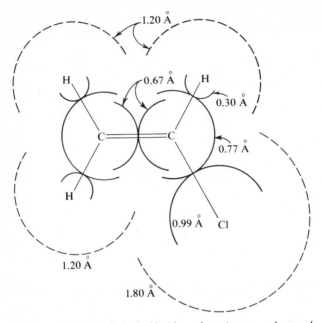

Fig. 1.2 *Geometry of vinyl chloride and pertinent covalent and van der Waals radii.*

to bond distances, these models also require information on bond angles (see the next section). For a more complete geometric description of ethanol it is also necessary to have information about preferred rotational angles at single bonds, discussed in Chap. 3.

BOND ANGLES

In addition to relatively fixed bond distances and bond strengths, another general feature of covalent bonds in polyatomic molecules is their relatively fixed geometry with respect to each other, which is generally expressed in terms of bond angles. Experimental information indicates that with a central atom bonded to two others, the geometry can be linear (bond angle = 180°) or bent (bond angles as small as 90°). With three atoms bonded to a central atom, the arrangement can be planar or trigonal (bond angles = 120°) or pyramidal (bond angles ≥ 90°).

The geometry of bond angles seems to be determined principally by the number of electrons in the valence shell of the central atom under consideration. For example, when there are only two pairs of bonding electrons, as in dicovalent compounds of group II elements, the least repulsion between electron pairs will occur when they are on opposite sides of the central atom, i.e., in a linear arrangement. The following are examples of linear molecules:

$H_3C-Hg-CH_3$ $|\bar{C}l-Be-\bar{C}l|$ $|\bar{C}l-Zn-\bar{C}l|$

Dimethylmercury Berylium chloride Zinc chloride

We shall use the convention of a line *between* atoms to represent a bonding pair of electrons, a line *alongside* atoms to represent a nonbonding pair of electrons.

For tricovalent compounds of group III elements, involving three pairs of bonding valence electrons, the minimum repulsion between electron pairs involves the trigonal bond angle of 120°, as in the following boron compounds:

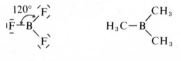

Boron fluoride   Trimethylboron

For carbon and other group IV elements, covalent bonding to four atoms involves tetrahedral geometry with bond angles of 109.5°, again the arrangement which minimizes electron-pair repulsions (Fig. 1.3). This same tetrahedral arrangement is found for CCl_4, SiF_4, and $NH_4^{\oplus}$, all similarly involving four atoms bound to a central atom by four electron pairs.

It is also possible to explain these geometric arrangements by looking at the *individual* electrons. For example, if the four valence electrons around beryllium were arranged tetrahedrally for minimum electron repulsion, this would also explain a linear

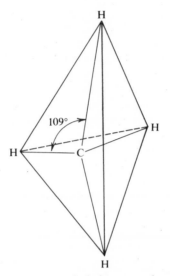

Fig. 1.3 *Tetrahedral geometry for methane.*

geometry in $BeCl_2$ (Fig. 1.4). Similarly an octahedral arrangement for six bonding electrons in BF_3 would lead to trigonal geometry (Fig. 1.5) and a cubic arrangement of eight valence electrons around carbon to a tetrahedral geometry. In the latter case, a square planar arrangement of nuclei would *not* minimize repulsions between the four protons as well as the nonplanar tetrahedral arrangement would.

When nonbonding electrons are attached to the central atom, some distortions from these simple geometries occur. For example, the bond angles in NH_3 (107.3°), PH_3 (93.3°), AsH_3 (91.8°), and SbH_3 (91.3°) diverge increasingly from the

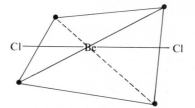

Fig. 1.4 *Tetrahedral arrangement of four bonding electrons in beryllium chloride; nuclei are linear.*

tetrahedral angle of 109.5° expected for four electron pairs. A similar trend is shown in the valence-bond angles for H_2O (104.5°), H_2S (92.2°), H_2Se (91.0°), and H_2Te (89.5°).

$$H-\bar{N} \begin{matrix} \\ | \\ H \end{matrix} {\scriptstyle\searrow} H$$

An explanation offered for this decreased valence angle, compared to 109.5°, is that the unshared electron pairs, confined by the positive electric field of only one atomic nucleus, occupy a greater volume than the shared pairs, confined by the larger positive field of the *two* atomic nuclei which share them. While

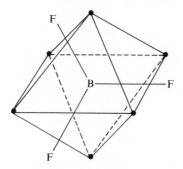

Fig. 1.5 *Octahedral arrangement of six bonding electrons in boron fluoride; nuclei are trigonal.*

there is no experimental way to "locate" the electrons, as x-ray can for nuclei, support for a larger volume occupied by the non-bonding electrons is indicated in the very much larger value for the van der Waals radii than the covalent radii (see Table 1.4). In any event, the argument is that the *nonbonding* pairs occupy a greater volume than the bonding pairs and thus distort the geometry, contracting the bond angle for the valence bonds.

An alternative explanation, which is equally satisfactory for bond angles, is based on orbital hybridization rather than simple concepts of electrical repulsion. For the case of Be, this view holds that the $2s$ and one $2p$ orbital combine to form two new equivalent sp hybrid orbitals. The values for the wave function ψ_{sp} for the two new orbitals are conceived as linear combinations of ψ_{2s} and ψ_{2p}, one being $\psi_{2s} + \psi_{2p}$, and the other $\psi_{2s} - \psi_{2p}$. In terms of orbital contours, this process can be represented as in Fig. 1.6.

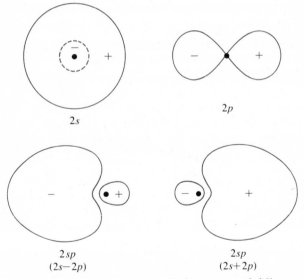

Fig. 1.6 *Representation of 2sp orbitals as sum and difference of 2s and 2p orbitals. The signs are for the wave function ψ.*

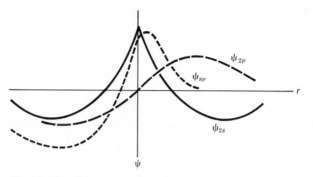

Fig. 1.7 *Plot of* ψ_{2s}, ψ_{2p}, *and* ψ_{2sp} *along the* 2_p *axis.*

The values of ψ and ψ^2 (the probability of finding the electron) along the axis of the p orbital are plotted in Figs. 1.7 and 1.8. It should be noted that the arithmetic to convert an s and a p orbital into two sp orbitals is completely reversible. Thus it is possible to construct two hybrid orbitals (s and p) by adding and subtracting two symmetrical sp orbitals.

In analogous ways, $2s$, $2p_x$, and $2p_y$ orbitals can be combined to form three new sp^2 hybrid orbitals, which will have trigonal geometry. It is thus equivalent, in referring to bond geometry, to describe it as trigonal or sp^2 geometry or, in the previous case, to describe it as linear or sp geometry.

In the same way, $2s$, $2p_x$, $2p_y$, and $2p_z$ orbitals can be combined to form four new equivalent sp^3 hybrid orbitals of tetrahedral (or sp^3) geometry.

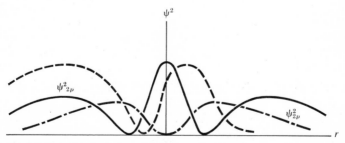

Fig. 1.8 *Plot of* $\psi_{2s}{}^2$, $\psi_{2p}{}^2$, *and* $\psi_{sp}{}^2$ *along the 2p axis.*

In order to explain the nearly right-angle bonds in AsH_3 and SbH_3 and H_2Se and H_2Te, it has been proposed that these compounds use *unhybridized* $4p$ or $5p$ orbitals for bonding. This geometry can thus be described as p^3 or p^2, respectively, for these group V and VI hydrides. On the basis of this hypothesis, the larger angle in NH_3 and H_2O has been explained as being due to a greater repulsion between the hydrogen atoms in these compounds since they are both closer together (smaller bond radii for N and O) and much more positively charged (much larger electron affinity for N and O).

Thus there are two contrasting explanations for the geometry at these atoms in groups V and VI: (1) they are generally of tetrahedral geometry, distorted by the larger volume occupied by the unshared pairs, a volume which increases as the electron affinity decreases going from N to Sb; or (2) they are of p^3 (or p^2) geometry, distorted by repulsion between charges on the hydrogen atoms which increase with increase in electron affinity of the central atom. We make no choice in interpretation, using either the geometrical or bond-hybrid terminology to indicate bond angles.

For pentacovalent structures with five electron pairs, the preferred geometry is a trigonal bipyramid, as shown for gaseous PCl_5:

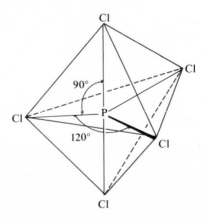

Three chlorine atoms are arranged trigonally with the fourth and fifth atoms above and below the center of the equilateral triangle defined by the first three. In terms of bond hybrids, this geometry may be described by five hybrids of $3s$, $3p_x$, $3p_y$, $3p_z$, and $3d_{z^2}$ orbitals, i.e., by five dsp^3 orbitals.

The geometries of SF_4 and ClF_3 represent compounds with unshared electron pairs among five pairs surrounding the central atom. The distortions observed from regular trigonal bipyramid geometry are in accord with the view that unshared pairs occupy more space than bonding pairs.

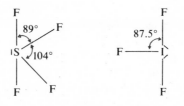

For six electron pairs, octahedral geometry is preferred. An example occurs in crystalline PCl_5, which has been shown to be an ionic lattice of PCl_4^+ cations (tetrahedral) and PCl_6^- anions (octahedral).

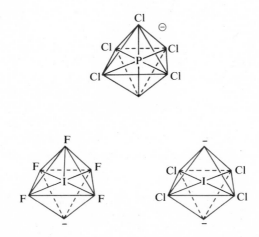

The geometry can also be constructed by hybridizing a $3s$, three $3p$, and two $3d$ orbitals to give six equivalent d^2sp^3 hybrid orbitals. Note that in the geometrical arrangements for the ICl_4 ion the unshared electron pairs occupy trans positions rather than adjacent sites. This is in accord with the idea that the unshared pair occupies a greater volume than a bonding pair.

GEOMETRY AT DOUBLE AND TRIPLE BONDS

So far we have discussed factors affecting bond angles involving compounds with single covalent bonds, but since many of the polymer molecules we shall discuss incorporate double as well as single bonds, we must consider geometrical factors involved in these bonds. We have already seen that such bonds are stronger (Table 1.2) and shorter (Tables 1.3 and 1.4) than corresponding single bonds. Their presence also has significant effects on other geometrical aspects of molecules.

For example, in ethylene, the H—C—H bond angles (116°) are significantly different from the usual tetrahedral angle of 109.5°:

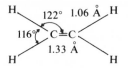

Furthermore, the barrier to rotation at the double bond is so substantial that the six atoms of ethylene are held in a planar arrangement, accounting for the fact that there are geometrical isomers of simple compounds like 1,2-dichloroethylene.

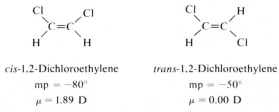

cis-1,2-Dichloroethylene
mp = −80°
μ = 1.89 D

trans-1,2-Dichloroethylene
mp = −50°
μ = 0.00 D

These are two distinct molecular species, with different physical and chemical properties. Which isomer is cis and which is trans was unequivocally determined by measurement of their dipole moment μ. The dipole moment, which we consider in more detail in Chap. 2, is a measure of electrical dissymmetry in a molecule. Since chlorine has a stronger electron affinity than carbon, the electron pair joining carbon to chlorine will spend more time near chlorine, thus giving chlorine a small net negative charge and carbon a small net positive one. In the trans isomer, the bond dipole moments thus created are equal and opposed, so that they would be expected to cancel, leading to no net dipole moment for this isomer. One isomer was indeed found to have no measurable dipole moment and thus can be unequivocally assigned the trans structure.

Two models have been proposed to represent the geometry of the C—C double bond. One starts with spheres with holes arranged at tetrahedral angles and joins the two atoms by two bent flexible "bonds" (actually springs instead of the rigid wooden sticks used to represent single bonds, see Fig. 1.9).

This model correctly represents a shorter internuclear distance (because the bonds are bent), restricts the rotation between the carbon atoms, and puts the six atomic nuclei in a plane. It can even explain the bond angles on the basis that the "bending" of the two bonds corresponds to a "compression" of the two electron pairs and permits the other two pairs to expand away from each other (from the tetrahedral 109.5° to the observed 116°).

The alternate explanation involves arguments based on hybrid orbitals. In this picture, the structural framework of ethylene is constructed from two trigonal carbons involving three sp^2 orbitals on each carbon for single bonds (σ bonds). This would predict 120° bond angles and would leave a p orbital on each carbon. Formation of a bonding hybrid orbital from overlap of these two orbitals can occur most effectively when the axes of the two orbitals are parallel. This concept envisions a shared

π-electron bond in this new bonding hybrid π orbital in which the second bonding pair will be above and below the plane defined by the four hydrogen atoms of ethylene.

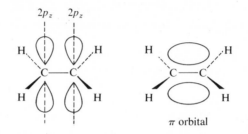

The bonding by the π orbital (64 kcal mole^{-1}) is not as great as that by the σ orbital (82 kcal mole^{-1}), but it will be destroyed entirely if one carbon is rotated so that the axis of the p orbitals involved are at right angles to each other. This feature also explains the large barrier to rotation at the C=C double bond, and the large net bond strength (146 kcal mole^{-1}) compared to the single bond can explain the contraction of the single bond (1.54 Å) to 1.33 Å when the extra 64 kcal mole^{-1} contribution to bond strength is exerted by the π bonding.

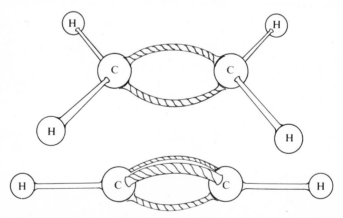

Fig. 1.9 *Molecular models of ethylene,* C_2H_4, *and acetylene,* C_2H_2.

Incidentally, two symmetrical hybrid orbitals can be constructed from the σ and π orbitals in much the same way that two sp orbitals were constructed from an s and a p orbital. The two new symmetrical bonding orbitals have a maximum electron density along lines corresponding to the bent bonds of the spring model in Fig. 1.9.

The triple bond in acetylene can be satisfactorily represented either by bent bonds (bottom drawing in Fig. 1.9) or by π orbitals (Fig. 1.10). The π orbital picture would involve a σ bond framework of linear sp orbitals at each carbon with $2p_y$ and $2p_z$ orbitals left free at each carbon, each of which can participate in π bonding in two new π orbitals. Since the sum of p_y and p_z orbitals is axially symmetrical around the x axis, little barrier to rotation would be expected at the triple bond.

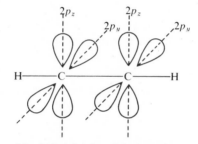

Fig. 1.10 *Orbital model of acetylene.*

Only a few molecules have two double bonds to the same carbon atom, examples being CO_2 and allene. The geometry of these arrangements is linear. The π bonds, however, are in planes at right angles to each other:

ASYMMETRIC (CHIRAL) MOLECULES

One of the consequences of bonds of fixed disposition in space is the possibility of making models which are mirror images of each other but *not* identical. It has recently been urged that the word *chiral* be used to represent this particular kind of asymmetry in molecules.

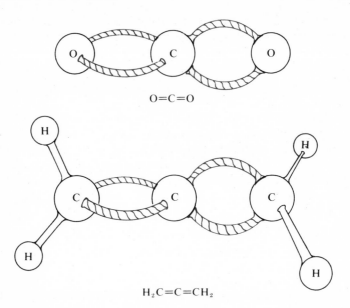

$O=C=O$

$H_2C=C=CH_2$

Fig. 1.11 *Molecular models of carbon dioxide*, CO_2, *and allene*, C_3H_4.

A tetrahedral atom will be chiral if there are four different groups attached to it. A classical example is lactic acid, which has been isolated in three different forms (one a mixture of the other two).

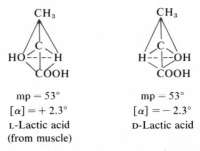

mp = 53°   mp = 53°
$[\alpha] = + 2.3°$ $[\alpha] = - 2.3°$
L-Lactic acid   D-Lactic acid
(from muscle)

That these two materials are different, despite the identity of most of their chemical and physical properties, is demonstrated by the fact that a 50:50 mixture has a different melting point (17°). The other evidence proving their difference is that one form rotates the plane of plane-polarized light clockwise, $[\alpha]_D = +2.3°$, while the other rotates it an equal amount in the opposite direction, $[\alpha]_D = -2.3°$. The 50:50 mixture, DL or racemic lactic acid, does not rotate the plane of polarized light; i.e., it is optically inactive.

There are many other examples of molecules whose models can be shown to be chiral and which have been isolated as D and L optically active forms. Properly substituted allenes and substituted sulfoxides and phosphines holding three different groups are examples.

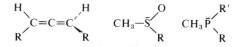

It is, of course, possible to construct models of an amine analogous to the phosphine above which is chiral. However, no such amine has ever been successfully separated into stable chiral isomers.

It has in fact been shown conclusively that for nitrogen the interconversion of these two forms is accomplished with great rapidity by the nitrogen atom passing through the plane of the other three groups. For phosphine and the elements lower in group V, which can all be isolated in stable optically active forms when properly substituted, it is presumably the sharper bond angles at the central atom (which puts it further out of the plane of the other three groups) and the greater mass of the atom which combine to prevent the interconversion by "turning the umbrella inside out."

PROBLEMS

1.1 From the data given in Table 1.4 estimate the interatomic distance in a nitrogen molecule (compare with the value in Table 1.3).

1.2 From the data given in Table 1.4, estimate the H-to-N distance in HC≡N (hydrogen cyanide). Compare this to the "closest" approach of H to N in two different HCN molecules.

1.3 Estimate the three different H-to-H distances in ethylene (using bond radii of Table 1.4 and assuming 120° bond angles).

SUGGESTED READING

Mahan, B.: "University Chemistry," chap. 11, Addison-Wesley Publishing Company, Inc., Reading, Mass., 1969.

Pauling, L.: "The Nature of the Chemical Bond," chaps. 4 and 7, Cornell University Press, Ithaca, N.Y., 1960.

2.
Intermolecular
Forces

INTRODUCTION

Since all atoms and molecules are in thermal motion, their kinetic energy being a direct measure of their temperature, one may ask: Why do atoms and molecules not fly apart? For molecules, the strong bond forces are so much greater than the kinetic energy at normal temperatures that the attractive bonding forces dominate. Of course, as one raises the temperature, the kinetic energy increases and ultimately will indeed overcome bonding forces. Under these conditions of high temperature, molecules become unstable and dissociate into fragments or atoms.

The same general considerations are involved in the aggregation of molecules to the liquid or solid states. These states are stable only when the forces between molecules are of sufficient magnitude to overcome the kinetic energy of the molecules.

DIPOLE MOMENTS AND BOND MOMENTS

One factor which can contribute to the attraction between molecules is the force exerted between dipolar charges in molecules. Let us consider, for example, the case of HCl or HF. In these molecules, the electron pair in the covalent bond is not equally shared between the nuclei, since chlorine and (especially) fluorine have much higher electron affinities than hydrogen. Because of this, the bonding electrons spend more time near the halogen atom in these molecules, leading to excess electron density there and some deficiency around the hydrogen. This can be expressed by saying there is a partial negative charge δ^- on the chlorine and a corresponding δ^+ on the hydrogen (since the molecule as a whole carries no charge, these *must* be equal charges, but opposite in sign):

$$H^{\delta+}—Cl^{\delta-}$$

Such a charge distribution in a molecule creates a *dipole moment*, $\mu = \delta d$, where δ is the charge (δ^+ and δ^-) so separated and d is the distance between δ^+ and δ^-. The electric dipole moment μ can be determined for molecules by measuring their response to imposed electric fields. The values for some typical molecules are summarized in Table 2.1. They are usually reported in Debye units (D) per molecule, which are equal to 10^{18} times the value in cgs units.

Table 2.1 Dipole Moments for Some Molecules

Molecule	μ, D	Molecule	μ, D
HF	1.98	H_2O	1.86
HCl	1.03	H_2S	1.1
HBr	0.79	H_2Se	0.4
CH_3F	1.81	NH_3	1.47
CH_3Cl	1.86	PH_3	0.55
CH_3Br	1.80	AsH_3	0.22
$H_2C{=}O$	2.50	$(CH_3)_3N$	0.75
CH_3OH	1.70	$CH_3C{\equiv}N$	3.9

The dipole moments for polyatomic molecules can be interpreted as arising from dipole moments associated with each bond, much as heats of formation can be ascribed to individual bond energies (Table 1.2). However, dipole moments are vector quantities, i.e., have both magnitude and direction, and so individual bond moments must be added vectorially to get the overall dipole moment of a molecule. The convention is to represent the direction of a dipole-moment vector by putting an arrowhead on the negative end and a cross at the positive end.

Let us consider water first. The overall dipole moment of 1.86 D must be made up of two equivalent bond moments associated with each H—O bond. The bond angle for water is 104.5°. This will then give a bond moment of 1.53 D associated with each H—O bond in water.

In a similar manner, the dipole moment of methyl chloride can be ascribed to bond moments associated with the C—Cl bond and the three C—H bonds. From many data on compounds with a C—H bond, the bond moment has been assigned

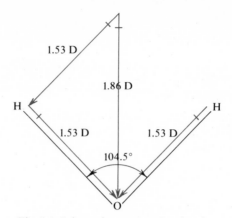

Fig. 2.1 *Scheme for representing the observed dipole moment of* 1.86 *D for water as the vector sum of two H—O bond moments of* 1.53 *D.*

as 0.3 D. Since methane has no dipole moment, the vector sum of three C—H bonds in a methyl group must be exactly equal to, collinear with, and opposite to the remaining C—H bond in methane and thus can easily be assigned the same value in CH_3Cl.

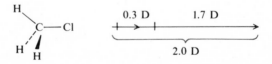

The overall dipole moment must thus be the resultant of a contribution of 0.3 D from the three C—H bonds and 1.7 D from the C—Cl bond.

The dipole moments of diatomic molecules, such as HCl, are normally interpreted by assuming that the distance d is equal to the internuclear separation, i.e., the bond length (see Table 2.3 or 2.4). This in turn leads to the possibility of assigning a value to $\delta^{\pm}$, the charge associated with the nuclei. The dipole moment for HCl (1.03 D) and the bond distance (1.27 Å) lead to a value of $+0.81 \times 10^{-10}$ esu on H and -0.81×10^{-10} esu on Cl:

$$\delta = \frac{\mu}{d} = \frac{1.03 \times 10^{-18} \text{ esu-cm}}{1.27 \times 10^{-8} \text{ cm}} = 0.81 \times 10^{-10} \text{ esu}$$

In the same way, charges can be calculated from bond moments to give an estimate of charges on each atom in more complex molecules. The values for water and methyl chloride are

$$\underset{\text{H}}{\overset{\text{H}-\text{O}}{\diagdown}} \begin{matrix} -3.16 \times 10^{-10} \\ +1.58 \times 10^{-10} \end{matrix} \qquad \underset{\text{H}}{\overset{\text{H}}{\diagdown}} \underset{\text{H}}{\overset{0.28 \times 10^{-10}}{\underset{0.13 \times 10^{-10}}{\text{C}}}} \text{Cl} -0.97 \times 10^{-10}$$

Table 2-2 summarizes average bond moments and related charges for a number of covalent bonds. Remember that the full electron charge is 4.8×10^{-10} esu, so that the central oxygen of water has two-thirds of an electron excess, the chlorine of methyl chloride about 20 percent of an electron excess.

Table 2.2 Average Bond Moments and Related Atomic Charges

Bond	μ, D	$d, \mathring{A}$	$\delta, esu \times 10^{10}$
(+) (−)			
H—C	0.30	1.08	0.28
C—O	0.85	1.43	0.60
C=O	2.40	1.24	1.93
C—S	0.95	1.81	0.53
C—N	0.40	1.47	0.27
C≡N	3.60	1.15	3.13
C—F	1.60	1.44	1.11
C—Cl	1.70	1.76	0.97
H—O	1.53	0.97	1.58
H—N	1.31	0.91	1.44
H—S	0.68	1.35	0.50

DIPOLE ATTRACTION

Given a molecule like HCl, HF, or H_2O, with strong dipoles, the attractive force between oppositely charged atoms can provide at least part of the cohesive force necessary to explain the existence of molecules in the liquid and solid states.

Let us consider the case of HCl. The mode of association of two HCl molecules which would result in maximum attraction schematically represented is

$$
\begin{array}{c}
3.0 \text{ Å} \\
\text{H} \text{-----} \text{Cl} \quad -0.81 \times 10^{-10} \\
1.27 \text{ Å} \\
\text{Cl} \text{-----} \text{H} \quad +0.81 \times 10^{-10}
\end{array}
$$

Each hydrogen has a charge of 0.81×10^{-10} esu; each chlorine -0.81×10^{-10} esu. The covalent-bond distance is 1.27 Å, and sum of van der Waals radii for H and Cl is 3.0 Å (Table 1.4). The new forces introduced by allowing two HCl molecules to approach as indicated above are two attractions between H and Cl (black dashed lines, $d = 3.0$ Å) and repulsions between two H's or two Cl's (red dashed lines, $d = 3.27$ Å). The net bonding attraction F will be the difference between these two (recall that the electrostatic force between any two charges is the product of the charges divided by the distance between them):

$$
F = \frac{(0.81 \times 10^{-10})^2}{3.0 \times 10^{-8}} - \frac{(0.81 \times 10^{-10})^2}{3.27 \times 10^{-8}}.
$$
$$
= 3.6 \times 10^{-14} \text{ erg molecule}^{-1}
$$

This value converts to 0.26 kcal mole^{-1}. Since the average kinetic energy of a molecule is $^3/_2RT$, at room temperature the kinetic energy (0.9 kcal mole^{-1}) is in considerable excess of the dipole attraction calculated. Actually, HCl is a gas boiling at $-85°$.

A similar calculation for HF, because of the smaller van der Waals radii (Table 1.4) and larger atomic charges (Table 2.2), leads to a calculated attractive force of 1.65 kcal mole^{-1}, and indeed HF is a liquid with a boiling point of 20°.

HEAT OF VAPORIZATION

One of the experimental measurements indicating the magnitude of attractive forces holding molecules together is the energy necessary to separate the molecules in the condensed phase in going to the gaseous state. Table 2.3 gives the experimental values for heats of vaporization of a number of simple molecules.

Table 2.3 Heats of Vaporization and the van der Waals Molecular-attraction Parameter a*

	ΔH_{vap}, *kcal mole*$^{-1}$	a, *liter*^{2} *atm mole*$^{-2}$
He	0.020	0.034
H_2	0.216	0.24
N_2	1.33	1.39
Ar	1.56	1.35
O_2	1.63	1.36
CH_4	1.95	2.25
Xe	3.02	4.19
CO_2	3.52	3.59
HCl	3.86	3.67
H_2S	4.48	4.43
HF	7.20	—
H_2O	10.52	5.46
$(CH_3)_2C{=}O$	7.65	13.91

*The parameter in the van der Waals equation of state, $(P + a/V^2)(V - b) = RT$, which corrects for attractive forces between gas molecules.

It has been found useful to express the attractive force measured by heats of vaporization as a parameter called the *cohesive-energy-density* parameter (CED), which is related to the heat of vaporization ΔH_{vap} and the molar volume of the liquid V.

$$CED = \frac{\Delta H_{vap}}{V} \text{ cal ml}^{-1}$$

One reason for this has been the observation that for molecular solutions, i.e., nonelectrolytes or nonionic solvents and solutes, the heat of mixing is proportional to the square of the difference in the square root of CED for solvent and solute:

$$\Delta H_{\text{mixing}} \propto \left[\left(\frac{\Delta H_{\text{vap}}}{V} \right)^{1/2} - \left(\frac{\Delta H'_{\text{vap}}}{V} \right)^{1/2} \right]^2$$

Values for CED are summarized in Table 2.4. The CED may be interpreted as a measure of the *polarity* of a molecule, at least of the degree to which its dipoles are available to interact with dipoles in other molecules. Since, as we saw earlier, electrical forces are the product of the interacting charges, the product of two $\sqrt{\text{CED}}$ parameters would be the proper form for translating *polarity* into an intermolecular force. The values of CED are also useful for predicting miscibility of liquids which produce molecular solutions without significant association or dissociation of molecules. The more positive the ΔH of mixing, the less

Table 2.4 Values for the CED

Compound	CED, cal ml^{-1}
CF$_4$	21
CH$_4$	23
Silicone rubber*	46
Heptane	55
CCl$_4$	74
Benzene	78
CHCl$_3$	88
CH$_2$Cl$_2$	94
Acetone	96
CS$_2$	102
CH$_3$CN	128
CH$_3$OH	198
H$_2$O	585

*Values for nonvolatile compounds, e. g., silicone rubber and other polymers, are conveniently estimated from the CED values for volatile solvents in which they show maximum solubility (see pages 77 and 90–92).

the liquids will tend to mix. If $\Delta H_{\text{mixing}} = 0$, entropy will dicate mixing, but as ΔH_{mixing} increases, the tendency toward incompatibility also increases. Normally, liquids with nearly equal CED will mix readily, while those with divergent CED will not.

ELECTROKINETIC VAN DER WAALS FORCES

While we have discussed electrostatic attractive forces between molecules arising from permanent dipoles, we can note in Table 2.3 that there are substantial attractive forces, as indicated by both ΔH_{vap} and a, for molecules such as CO_2 and CH_4 which have no electrostatic dipole moment. The explanation offered for these molecules is that while there is no permanent polarization, the electrons are in motion and are polarizable. The picture for a hydrogen molecule may be that at one moment in time the electrons are nearer one hydrogen and at another nearer the other, although the time-average distribution is symmetrical.

$$\overset{\delta+}{H}\!\!\!-\!\!\!\overset{\delta-}{H} \;\rightleftharpoons\; \overset{\delta-}{H}\!\!\!-\!\!\!\overset{\delta+}{H}$$

Another hydrogen molecule nearby could also have electrons moving at the same velocity, and, in fact, the temporary polarity in one would tend to polarize the other so as to engender a favorable attractive interaction between the molecules.

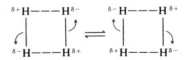

Such a synchronized polarization would lead to attractions indicated by the dashed lines and is termed *electrokinetic attraction*.

Even for the monatomic gases, such polarization can occur.

$$\overset{\delta-}{:}\overset{\delta+}{He}\;\overset{\delta-}{:}\overset{\delta+}{He} \;\rightleftharpoons\; \overset{\delta+}{He}\;\overset{\delta-}{:}\overset{\delta+}{He}\;\overset{\delta-}{:}$$

In view of the high velocities of electrons, the rate of such changes in polarity would be extremely high, of the order of magnitude of 10^{16} sec^{-1}!

The magnitude of these electrokinetic interactions will increase with the number of electrons and also with their polarizability. This trend can be seen in the values for intermolecular forces between the molecules in Table 2.3, excluding those which have attraction due to fixed dipole moments (H_2O, HF, HCl).

VAN DER WAALS REPULSIONS

The van der Waals attractions between molecules can be ascribed to electrostatic and electrokinetic forces. But what are the repulsive forces which establish a limit to the close approach of one molecule to another, i.e., which define the van der Waals radii of atoms and molecules (see Table 1.4 and Fig. 3.1)? The quantum theory of atomic and molecular structures predicts that an electron in an atom or molecule is in very rapid motion and that we can accurately predict only its average distribution. The geometrical shape of the electron distribution is referred to as an *orbital* and the geometric function which describes it as a *wave function* ψ.

For a hydrogen molecule, the contours of decreasing electron probability ψ^2 for the orbital of the bonding pair are represented in Fig. 2.2. These contours represent surfaces (by rotation around the bond axis) of equal electron probability (or density), high near the two nuclei and decreasing with distance. These orbital contours predict that while the highest electron density is near the nuclei and between them, there is considerable electron probability on the periphery of the molecule. It is the strong repulsions generated when electrons in these nonbonding orbitals interpenetrate that cause the sharp rise in energy (see Fig. 3.1) when molecules approach closer than the van der Waals radii established by these orbital dimensions.

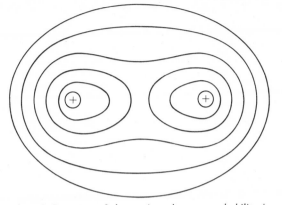

Fig. 2.2 *Contours of decreasing electron probability in the hydrogen molecule*, H_2.

PROBLEMS

1.1 On a drawing for the structure of formaldehyde, $H_2C = O$, indicate likely bond angles and distances.

1.2 Use this molecular geometry and the dipole moment of formaldehyde, $\mu = 2.50$ D, and bond moment for the CH bond, 0.3 D, to estimate the bond moment for the $>C=O$ bond.

1.3 From van der Waals radii in Table 2.4, estimate the maximum coulomb attraction between two formaldehyde molecules.

1.4 Given the CED for acetonitrile, CH_3CN, as 128 cal ml^{-1} and the density of the liquid near the boiling point as 0.75, estimate the ΔH_{vap}.

SUGGESTED READING

Mahan, B.: "University Chemistry," chap. 11, Addison-Wesley Publishing Company, Inc., Reading, Mass., 1969.

Pauling, L.: "Nature of the Chemical Bond," chaps. 4 and 7, Cornell University Press, Ithaca, N.Y., 1960.

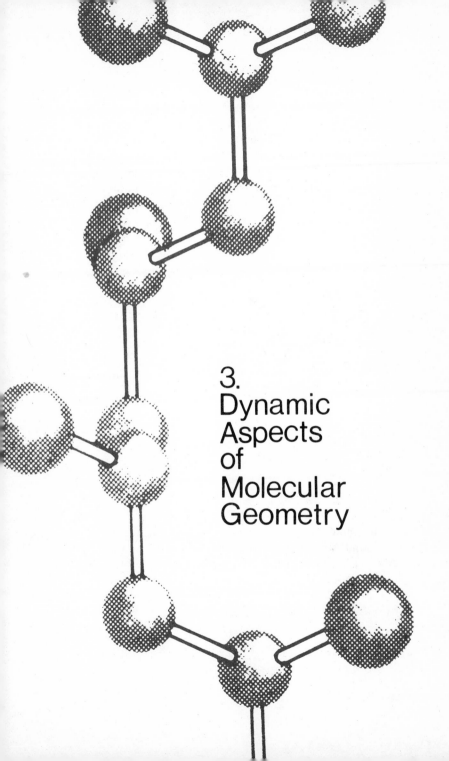

3.
Dynamic
Aspects
of
Molecular
Geometry

BOND STRETCHING

Since all molecules possess kinetic energy proportional to the
ambient temperature, it is important to consider the influence of
this motion on the geometry of molecules. One effect of such
motion is on bond lengths. Since equilibrium bond distance is
a balance between repulsive forces, which dominate at shorter
distances, and attractive forces, which dominate at longer
distances (see Fig. 3.1), thermal kinetic energy in a molecule

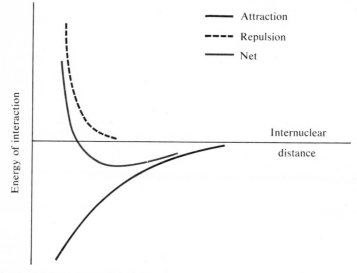

Fig. 3.1 *Interatomic forces.*

causes bonds to stretch and contract with a frequency related to
the bond strength and the effective mass at each end of the
bond. A very satisfactory approximation of this behavior is
the harmonic oscillation of two weights attached by a spring.

For such harmonic oscillators, this relationship is expressed
by

$$4\pi^2\nu^2 = \frac{k}{M_r} \tag{1}$$

In this expression ν is the frequency, i.e., the number of oscillations per second, and k is a proportionality constant, called the *force constant* of the spring, characteristic of the *restoring force* of the particular oscillator. The *reduced mass*, M_r, is defined by

$$\frac{1}{M_r} = \frac{1}{M_1} + \frac{1}{M_2}$$

In the simple case of one end of the spring being fixed, it is the mass; when the two weights are equal, $M_r = M/2$.

For any oscillator of this kind it should be recognized that there is a continuous interconversion of kinetic and potential energy. At the equilibrium separation, no potential energy is associated with the spring. At the maximum or minimum separation, the masses are momentarily at rest and therefore have no kinetic energy. Thus, there is a continuous interconversion of kinetic and potential energy in such an oscillator, with all the energy kinetic at the equilibrium separation, all potential at the extremes. At the extreme displacement ΔX, the potential energy (PE) is also related to the force constant k by

$$2(\text{PE}) = k \, \Delta x^2 \tag{2}$$

Equations (1) and (2) can be interrelated since each contains the force constant k

$$k = 4 \pi^2 \nu^2 M_r = \frac{2(\text{PE})}{\Delta x^2} \tag{3}$$

For typical molecules, the frequency of oscillations according to Eq. (1) is observable in infrared and Raman spectra. From such experimental observations of ν and assignment of M_r based on known molecular structure, values for k can be calculated. Values for force constants derived from experimental data in this way are summarized in Table 3.1. Note that for the carbon-carbon bond there is a correlation between k and bond strength (Tables 2.3 and 2.4). As the bond strength increases in the double and triple bond, so does the force constant k.

Table 3.1 Values for Force Constants k for Bond Stretching

Bond	k, dyne-cm $\times 10^{-5}$	Bond	k, dyne-cm $\times 10^{-5}$
C—H	5.0	C≡N	18.1
C—C	4.5	N—H	6.5
C=C	9.8	HF	9.7
C≡C	15.6	HCl	5.2
O—H	7.6	SH	4.0
C—O	5.4	PH	3.1
C=O	12.3	SiH	2.7
C—N	5.2		

On the basis of Eq. (2), it is possible to estimate the magnitude of Δx, the maximum deviation from the equilibrium bond distance. To do this we need to know the potential energy at maximum displacement. Since on the average the kinetic energy and potential energy of such an oscillator are equal, we can relate the potential energy at maximum displacement to thermal kinetic energy, $3RT/2$. Let us take the force constant of 4.5×10^5 dyne-cm for the carbon-carbon single bond and make this calculation for ethane at 300 K. For the purposes of this calculation, we may assume that the masses in motion are the entire methyl group, so that $M_r = {}^{15}/_2$ g mole^{-1}. In correct units, we must express the thermal energy in dyne-centimeters rather than kilocalories, and so for this purpose $R = 8.314 \times 10^7$ dyne-cm mole^{-1} K^{-1}, or for each molecule, $R = 1.385 \times 10^{-16}$ dyne-cm K^{-1}. Then

$$\Delta x = \left(\frac{3RT}{k} \right)^{1/2} = \left(\frac{3 \times 300 \times 1.385 \times 10^{-16}}{4.5 \times 10^5} \text{ cm}^2 \right)^{1/2}$$
$$= 4.5 \times 10^{-10} \text{ cm or } 0.045 \text{ Å}$$

Thus this bond, with an equilibrium length of 1.54 Å, oscillates at room temperature between lengths of 1.585 and 1.495 Å. Note that an increase in temperature will increase this extent of oscillation and an increase in the force constant will decrease it.

While frequencies of oscillation can be observed experimentally, e.g., by infrared and Raman spectroscopy, they can also be estimated from force constants and structure of molecules by Eq. (1), remembering that 1 dyne = 1 g-cm sec^{-2} and that M_r is per molecule. This gives

$$\nu = \left(\frac{k}{4\pi^2 M_r}\right)^{1/2} = \left(\frac{4.5 \times 10^5 \times 6.02 \times 10^{23}}{4\pi^2 7.5} \text{ sec}^{-2}\right)^{1/2}$$
$$= 3.02 \times 10^{12} \text{ sec}^{-1}$$

Note that this frequency of thermal oscillation will increase with the square root of the force constant and decrease inversely with the square root of the mass. For most covalent bonds, the frequency is near 10^{12} oscillations per second.

The frequency of oscillations of molecules in liquids or solids, with respect to their neighbors, is somewhat lower. The strength of van der Waals bonding holding molecules together in the condensed states may be of the order of one-tenth that of covalent bonding and for molecular units of about 100 amu in motion, the frequency of oscillation will then be about 10^{11} sec^{-1}. In this case this is equivalent to the number of "collisions" between a pair of molecules, but since a molecule is, on the average, surrounded by six nearest-neighbor molecules, the total number of collisions a molecule undergoes is also close to 10^{12} sec^{-1}.

BOND BENDING

In molecules with more than two atoms, another form of dynamic motion is that which will deform bond angles from their preferred equilibrium angle. For a water molecule, for example, the observed angle of 104.5° can be increased or decreased by bending. In the same way that bond stretching can be treated as a harmonic oscillation, so can bond bending, by equations analogous to Eqs. (1) and (2). The frequency will be related to the "stiffness" of the bond (i.e., to a bending force constant which measures its tendency to resist bending) and to the mass of the atoms undergoing motion in bending. Increase in bond stiffness will increase frequency, while increase in mass will decrease frequency. Furthermore, there is a similar interconversion of kinetic and potential energies, with no potential energy at the equilibrium angle and no kinetic energy at the maximum bending displacement.

The observed frequencies for most bond bendings are about the same as for stretching and are observable in infrared and Raman spectra in the range near 10^{12} sec^{-1}.

The potential energy for displacement from equilibrium molecular geometry by either bond stretching or bending is related to the square of the displacement [Eq. (2)]. An important consequence of this relationship is that a molecule responds to a force tending to distort it, e.g., a molecular collision, by adjusting as many stretching and bending modes as possible. We may illustrate by a simple example of a linear molecule, A—B—A. If each bond can expand and contract 0.05 Å at an energy input of 600 cal mole^{-1}, then the normal expansion and contraction of ABA will be 0.1 Å with an energy of 1200 cal mole^{-1}. If all the 0.1 Å were concentrated in *one* bond, it would require a fourfold increase in the energy in this bond, or 2400 cal mole^{-1}. Suppose this molecule is stressed so that it must contract by 0.2 Å (rather than 0.1 Å). This could be accommodated by having all the contraction occur in one bond. According to Eq. (2), increasing Δx fourfold will require a sixteenfold increase in energy, or in this case 9600 cal mole^{-1}. The same Δx could also be achieved by contracting each bond by 0.1 Å, which would require only 4800 cal mole^{-1}. It is thus a general principle of molecular motion, as a consequence of Eq. (2), that whenever possible molecular deformation will occur by a sum of small deformations of many bonds rather than a large deformation of one.

QUANTIZATION OF MOLECULAR VIBRATIONS

For an oscillator, quantum theory postulates that allowed energy states will differ by a constant energy $h\nu$, where ν is the frequency of the oscillator and h (Planck's constant) is 6.62×10^{-27} erg-sec. Since the frequency of oscillation of atoms in solids and in molecules is near 10^{12} sec^{-1}, this relationship predicts that the quantum gap between energy levels for atomic oscillations in molecules (and crystals) is about 6.62×10^{-15} erg atom^{-1}. In other words, the harmonic oscillators in molecules cannot have any energy whatsoever but only energies differing by the value $h\nu$. Since 1 erg $= 2.389 \times 10^{-8}$ cal, 6.62×10^{-15} erg atom^{-1} corresponds to 95 cal mole^{-1}. With average thermal kinetic energy at room temperature of about 600 cal mole^{-1}, this is quite sufficient to excite bending and stretching modes in molecules. We therefore must expect the geometry of molecules to be dynamic, with considerable excitation of bending and stretching modes of vibration, i.e., with some molecules in the unexcited, or ground, state and others containing $h\nu$, $2h\nu$, or $3h\nu$, etc., excess energy.

ROTATION AT SINGLE BONDS

Another important motion of molecules is rotation, which occurs at single bonds. One of the simplest molecules in which to consider the consequences of this motion is ethane. At one time it was believed that there was "free" rotation of one methyl group with respect to the other in ethane, C_2H_6. Pitzer, however, found that this hypothesis did not adequately explain heat-capacity data for ethane and proposed that there must be a barrier of about 3 kcal mole^{-1} for this rotation and that the energy maximum occurs at what is called the *eclipsed* conformation, the minimum at the *staggered* conformation. *Conformation* has become the accepted term to refer to the geometrical changes in shape of molecules due to rotational changes at single bonds.

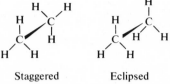

Staggered   Eclipsed

The change in energy with rotation can be represented by Fig. 3.2, which plots the energy of the system as a function of rotation of the single bond. The angle of rotation (dihedral angle) for one eclipsed form is assigned 0°.

The barrier to interconversion from one staggered conformation to another is about 3 kcal mole^{-1}, and the frequency of such interconversion is about 10^4 sec^{-1} at room temperature. The oscillatory vibrations corresponding to the red energy levels (separated by about 100 cal mole^{-1}) occur with a frequency of about 10^{12} sec^{-1}. The concept of rotational motion is therefore

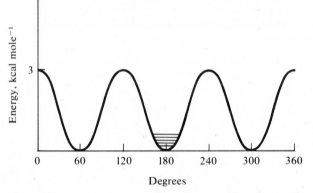

Fig. 3.2 *Variation in energy with conformation for ethane. The dihedral angles of* 60, 180, *and* 300° *correspond to staggered conformations, of* 0, 120, *and* 240° *to eclipsed conformations. The energy levels indicated in color for the* 180° *conformation correspond to thermally populated energy levels for harmonic oscillatory rotations around the* 180° *conformation. There will be similar levels for the other two staggered conformations, at* 60 *and* 300°.

one involving very rapid (about 10^{12} sec^{-1}) rotatory oscillations around the stable staggered conformation with occasional (about 10^4 sec^{-1}) rotations over the barrier to another stable conformation.

Values for barriers to rotation have been estimated from a variety of experimental data. Some typical examples are summarized in Table 3.2. Note that the energy holding these molecules in a preferred conformation is of the same order of magnitude as the van der Waals forces which hold liquid molecules together, as measured by heats of vaporization (Table 2.3).

Table 3.2 Estimated Barriers to Rotation

Molecule	kcal mole^{-1}	Molecule	kcal mole^{-1}
CH_3-CH_3	3.0	$CH_3-CH_2CH_3$	3.3
CH_3-CH_2F	3.30	$CH_3-CH(CH_3)_2$	3.9
CH_3-CH_2Cl	3.56	$CH_3-C(CH_3)_3$	4.8
CH_3-OH	1.07	Cl_3C-CCl_3	10.8
CH_3-OCH_3	2.7	$Cl_3Si-SiCl_3$	1.0
CH_3-SiH_3	1.70	CH_3-SH	1.26
CH_3-GeH_3	1.2	CH_3-SCH_3	2.13
CH_3-NH_2	1.9		

The exact nature of the barriers to rotation is not a matter of agreement. No doubt one factor is an increase in van der Waals repulsions when nonbonded atoms are brought closer together in the eclipsed form. Such a view is supported by the increase in the barrier for ethane (3.0 kcal mole^{-1}) when hydrogen atoms in one methyl group are replaced by fluorine (3.3), chlorine (3.56), methyl (3.3), two methyls (3.9), or three methyls (4.8). The marked decrease in barrier between C_2Cl_6 (10.8) and Si_2Cl_6 (1.0) is also consistent with this view, since the increased bond length for the silicon-silicon bond (2.34 Å) compared to carbon-carbon (1.54 Å) would greatly increase the minimum chlorine-chlorine distance in the eclipsed form. For C_2Cl_6 in the eclipsed form, the chlorine-chlorine distance is 2.75 Å, compared to the sum of van der Waals radii (Table 1.4) of 3.6 Å, while in Si_2Cl_6, the chlorine-chlorine distance in the eclipsed form is 3.77 Å, just slightly greater than the minimum van der Waals separation of 3.6 Å.

All the compounds in Table 3.2 have a conformational energy diagram similar to that of ethane (Fig. 3.2) with a threefold axis of symmetry. 1,2-Dichloroethane, however, presents a new feature, since the various eclipsed and staggered forms are no longer identical. The three staggered forms are represented below in two different ways:

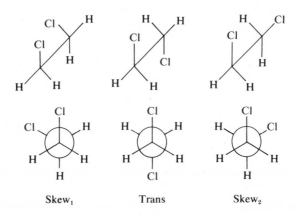

Skew₁   Trans   Skew₂

The conformation with the chlorines opposed is called the *trans* conformation, while the other two are called *skew* or *gauche* conformations. We shall prefer *skew* as it is a more accurately descriptive term. (Gauche comes from the French meaning "left"; actually the skew forms can be turned either right or left, and the models correspond to chiral models, i.e., D and L forms.)

Similar considerations apply to butane, C_4H_{12}.

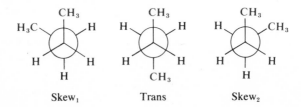

Skew₁   Trans   Skew₂

The energy diagram for conformations in butane is given in Fig. 3.3.

The energy difference between trans and skew forms for butane is 0.9 kcal mole^{-1}, while the barrier between trans and either skew form, $\ddagger_1$, is 3.5 kcal mole^{-1}. The barrier between the two skew forms, $\ddagger_2$, which places the two methyl groups in an eclipsed conformation, is 4.5 to 6.1 kcal mole^{-1}. Models show that in the conformation with methyls eclipsed, the distance between nonbonded hydrogen atoms in the two methyl groups is a minimum of 1.85 Å, while the sum of the van der Waals radii for two hydrogen atoms is 2.4 Å. There is thus considerable interpenetration of nonbonded hydrogens in this form, accounting for its higher energy.

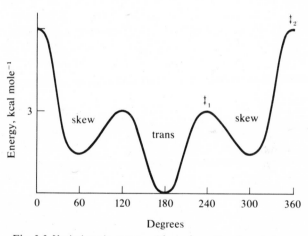

Fig. 3.3 *Variations in energy with conformation for n-butane.*

From an estimate of the energy difference between trans and skew forms, it is possible to calculate the relative population of the conformations. We may represent the number of molecules in the skew$_1$ conformation as N_{sk} (there will be an equal number of skew$_2$ since the two skew forms are equal in energy) and the number of molecules in the trans conformation as N_{tr}.

$$K = \frac{2N_{sk}}{N_{tr}} = \exp\left(\frac{-\Delta F}{RT}\right) \tag{4}$$

In order to evaluate this equation we assume that the entropy differences for skew and trans forms are negligible so that ΔF ($= \Delta H - T\,\Delta S$) may be approximated as ΔH. For the cases of butane and 1,2-dichloroethane ($\Delta H \simeq 1$ kcal mole^{-1}), this calculation predicts that about 72 percent of the molecules will be in the trans form at room temperature with 14 percent skew$_1$ and 14 percent skew$_2$.

For the case of 1,1-difluoro-1,2-dibromo-2,2-dichloroethane, the rate of interconversion of trans and skew forms has been estimated experimentally by the nuclear magnetic resonance spectra of ^{19}F. The two forms have different spectra because the fluorine atoms are in different environments in the different conformations. It turns out that the trans-to-skew ratio is 1.4 : 1 at $-80°C$ and the rate of interconversion of conformations is much less than 200 sec^{-1} at $-80°C$ but considerably faster than 200 sec^{-1} at 25°C.*

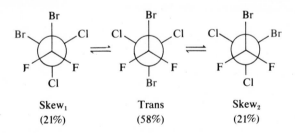

Skew$_1$ Trans   Skew$_2$
(21%)   (58%)   (21%)

CYCLOHEXANE CONFORMERS

Examples of important and widely studied cases of conformational equilibria and interconversion are various derivatives of

---

*J. D. Roberts, "Nuclear Magnetic Resononce," McGraw-Hill Book Company, New York, 1959.

cyclohexane compounds, including the biologically important steroid hormones. For cyclohexane it has long been recognized that two different strainless conformers can be prepared using conventional ball and peg models.

It turns out that the so-called *chair form* is a *rigid* structure when made by ball and peg models, while the so-called *boat* form is *flexible* and can be rotated easily so that C_1 and C_4 or C_2 and C_5 or C_3 and C_6 are the "bow" and "stern" of the boat, i.e., are the two atoms above the plane of the other four.

Furthermore, examination of the models (Fig. 3.4) reveals that all six carbon-carbon bonds in the chair form are in the preferred staggered conformation, whereas two of the six bonds in the boat form are in the unfavorable eclipsed form. Best experimental estimates of the relative stabilities of the boat and chair form for cyclohexane indicate the boat form to be less stable (higher in energy) by 5.6 kcal mole^{-1}. This corresponds to an increase of energy of 2.8 kcal mole^{-1} for each of the two

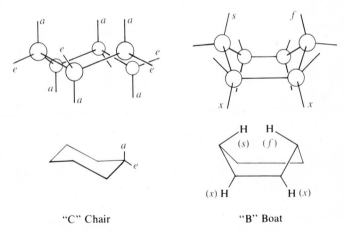

"C" Chair   "B" Boat

a Axial bonds   *s-f* Bowsprit-flagpole interaction
e Equatorial bonds   *x-x* Pair of eclipsed bonds

Fig. 3.4 *Chair (C) and boat (B) forms of cyclohexane. (Redrawn from Ernest L. Eliel, "Stereochemistry of Carbon Compounds," p. 205, McGraw-Hill Book Company, New York, 1962.)*

eclipsed bonds in the boat form, which is in reasonably close agreement with the 3 kcal mole^{-1} higher energy for the eclipsed form of ethane compared to the more stable staggered form.

Note that in the more stable chair conformation of the cyclohexane ring, there is a set of six hydrogen atoms (one on each carbon) directed outward from the ring, while the other hydrogens are alternately directed above or below the ring (see Fig. 3.5). The former bonds are referred to as *equatorial* and the latter as *axial*. The internuclear distance for hydrogen atoms in axial positions is 2.54 Å (appreciably longer than the sum of van der Waals radii for nonbonded hydrogen atoms of 2.4 Å), so that there is evidently no significant contribution from van der Waals repulsions of axial hydrogens to the energy of the boat form. However, if a group with a larger van der Waals radius than hydrogen is substituted for hydrogen, one may ask: What is the in-

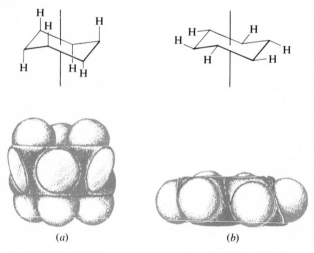

(a) (b)

Axial hydrogens Equatorial hydrogens

Fig. 3.5 (a) *Axial hydrogens*; (b) *equatorial hydrogens*. (*Redrawn from Melvin S. Newman, "Steric Effects in Organic Chemistry," pp. 16 and 17, John Wiley & Sons, Inc., New York, 1956.*)

fluence of the size of this group on the relative stability of the axial and equatorial conformers, and on the rate of interconversion of the two forms?

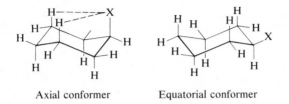

Axial conformer   Equatorial conformer

Since the rate of interconversion has been shown to be rapid at room temperature, it is not possible to isolate such conformers as stable isomers, but the question of their equilibrium concentration for various groups X has been extensively investigated. The equatorial isomer, as expected, has proved to be the more stable form, and, generally speaking, the larger the size of X (the greater the van der Waals repulsions between X and the two adjacent hydrogens indicated by the red dashed lines), the greater the energy factor favoring the equatorial conformer.

Figure 3.6 represents the energy diagram for conformational changes in methylcyclohexane, and Fig. 3.7 shows models

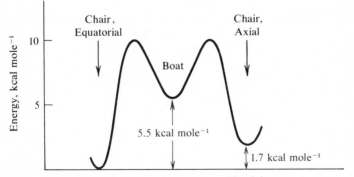

Fig. 3.6 *Conformational energy changes in methlcyclohexane.*

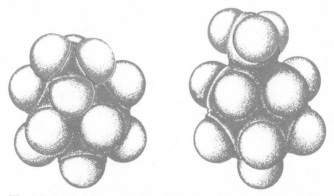

Fig. 3.7 *Axial and equatorial methyl groups. (Redrawn from Melvin S. Newman, "Steric Effects in Organic Chemistry," pp. 16 and 17, John Wiley & Sons, Inc., New York, 1956.)*

of the axial and equatorial conformers. Table 3.3 summarizes data for a number of substituents.

Note that these energy differences would correspond to the energy differences expected for a methyl and X in the skew form

Table 3.3 Conformational Energy Favoring the Equatorial Conformer for Monosubstituted Cyclohexanes*

X	ΔF, $kcal\ mole^{-1}$
OH	0.4 - 0.9
OAc	0.4 - 0.7
OCH_3	0.5 - 0.7
F	0.25
Cl	0.3 - 0.5
Br	0.2 - 0.7
CH_3	1.5 - 1.9
C_2H_5	2.1
C_6H_5	2.6
t-Bu	>4.4

*Data from E. L. Eliel, "Stereochemistry of Carbon Compounds," p. 236, McGraw-Hill Book Company, New York, 1962.

(axial) vs. the trans form (equatorial).

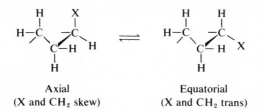

Axial	Equatorial
(X and CH_2 skew)	(X and CH_2 trans)

Recently, Eliel and Knoeber* have extended work on conformational analysis to six-membered rings containing oxygen, which are of biological significance since most glucose units in carbohydrates exist in this form.

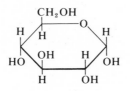

α-Glucose

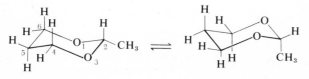

2-Methyl-1,3-dioxane

The conformational energy differences, summarized in Table 3.4, are similar to those for the cyclohexane ring itself.

Note that the methyl group in the 2-position of 1,3-dioxane has a greater energy in the axial position than in cyclohexane. This could be due in part to the shorter bond length for C—O than for C—C bonds as well as to the sharper bond angle at oxygen. Both these factors would place the axial 2-methyl group

*J. Amer. Chem. Soc., **90**:3444 (1968).

Table 3.4 Conformational Energies Favoring the Equatorial Conformer in Substituted 1,3-Dioxanes*

Substituent	ΔF, kcal mole^{-1}
2 - Me	3.55
4 - Me	2.9
5 - Me	0.8
5 - Et	0.7
5 - i - Pr	1.0
5 - C_6H_5	1.0
5 - t - Bu	1.4

*Eliel and Knoeber, *J. Amer. Chem. Soc.*, **90**:3444 (1968).

closer to the two axial hydrogens in 1,3-dioxane than in cyclohexane. In the 5-position, however, the axial energy in 1,3-dioxane is less than for cyclohexane since in this instance there are *no* opposing axial hydrogens.

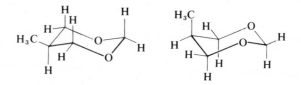

CYCLOPENTANE CONFORMERS

For many years cyclopentane was assumed to be a flat planar ring system, since the internal angle for a regular pentagon (108°) is less than the tetrahedral angle of 109.5°. It has now become evident that the five-membered ring is also puckered out of the plane and can exist in two preferred conformations, the *envelope* and the *half chair*.

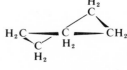

Planar   "Envelope"   "Half chair"

The driving force for puckering this ring system out of a planar arrangement, which would be preferred on the basis of carbon bond angles, is believed to be relief of the five eclipsed conformations that would exist in the planar form. The half envelope would have only three eclipsed bonds and the half chair only two. These forms would thus benefit from relief of about 6.0 and 9.0 kcal mole^{-1} of eclipsed conformational strain energy. These considerations are of biological significance since a number of five-membered rings exist in important biological molecules, e.g., the ribose and deoxyribose units in ribonucleic acid (RNA) and deoxyribonucleic acid (DNA), respectively.

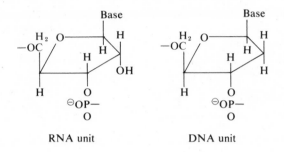

RNA unit   DNA unit

CONFORMATIONS AT O—O AND S—S BONDS

Two other biologically important bonds for which conformations are significant are the oxygen-oxygen bond in peroxides, e.g., hydrogen peroxide, H_2O_2, and the sulfur-sulfur bond in disulfides, e.g., cystine or thioctic acid.

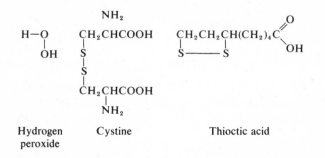

Hydrogen   Cystine   Thioctic acid
peroxide

It turns out that these molecules prefer a conformation with the two covalent bonds forming a 90° dihedral angle.

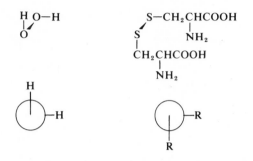

For dimethyl disulfide ($R = CH_3$), the dihedral angle has been found to be 90° for the preferred conformer, while the C—S—S angle is 107°. The barrier to rotation at the S—S bond is 6.8 kcal mole^{-1}, very much larger than the 1.5 kcal mole^{-1} barrier at the C—S bond in the same molecule.

There has been speculation about the reason for the skewed conformation of these molecules. One proposal interprets the geometry in terms of atomic orbitals. This hypothesis assumes that at divalent oxygen or sulfur, the two covalent-bonding electron pairs utilize $2p$ (or $3p$) orbitals, accounting for the bond angle close to 90° at oxygen and sulfur. This hypothesis would further postulate that the two *nonbonding* pairs at each oxygen (or sulfur) would then occupy an s and a p orbital. The electrons in the spherically symmetrical s orbitals would impose no preference for restricted rotation at O—O or S—S bonds. The *nonbonding* unshared pairs occupying p orbitals, however, would have a minimum repulsive interaction if the axes of these p orbitals were at right angles to each other.

For those preferring the view of tetrahedral geometry for four valence-shell electron pairs around atoms such as oxygen or sulfur, one can also rationalize the observed geometry. In this case, minimizing repulsion of unshared *nonbonding* pairs would predict a dihedral angle of 60° between the O—H bonds. To

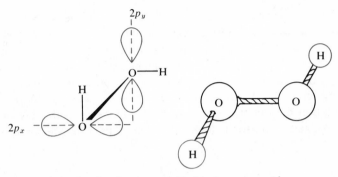

Fig. 3.8 *Orbital and molecular models of hydrogen peroxide.*

explain the 90° angle observed, one must invoke electrical or van der Waals repulsion between the groups attached to oxygen or sulfur to expand the predicted 60° angle to the observed 90°.

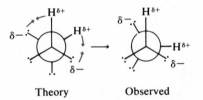

Theory   Observed

One interesting and significant consequence of the preferred 90° dihedral angle at sulfur-sulfur bonds is the unusual properties conferred on five-membered cyclic disulfides such as the biologically important molecule thioctic acid. The geometry of a five-membered ring would tend to impose a planar geometry, which would correspond to a 0° dihedral angle at S—S bond. While the ring is undoubtedly *not* planar,

it also seems evident from models that it would require very large bond-angle or bond-distance deformations to accommodate a 90° dihedral angle at the S—S bond.

The longer-wavelength light absorbance at the S—S bond in thioctic acid, $\lambda_{max} = 330$ nm compared to $\lambda_{max} = 250$ nm for unstrained disulfides, e.g., cystine or cyclic disulfides with larger rings (see Fig. 3.9), is one consequence of this strain. Furthermore, the excited molecule after absorption of light undergoes ready scission of the S—S bond. This scission occurs after every excitation for thioctic acid (quantum yield $= 1.0$) but only once in 50 excitations for a linear disulfide (quantum yield $= 0.02$).

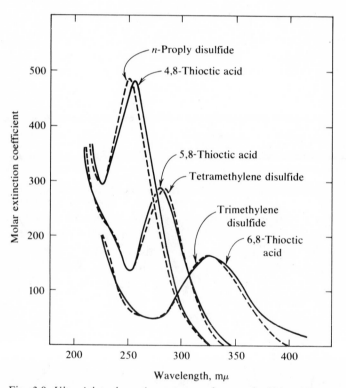

Fig. 3.9 *Ultraviolet absorption spectra of some disulfides.* (*From J. A. Baltrop, Hayes, and Calvin,* "*The Chemistry of 1,2-Dithiolane (Trimethylene Disulfide) As a Model for the Primary Quantum Conversion Act in Photosynthesis, J. Amer. Chem. Soc.,* **76**: 4348 (1954).

The biological functions of thioctic acid are related to this photodissociation, and thus the biological function of thioctic acid is based on the strained dihedral angle in the molecule.

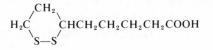

6,8-Thioctic acid
(naturally occurring)

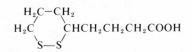

5,8-Thioctic acid
(6-membered ring)

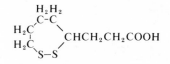

4,8-Thioctic acid
(7-membered ring)

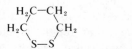

Tetramethylene disulfide

Trimethylene disulfide

One other interesting feature of geometry at the S—S bond is revealed in the ultraviolet spectrum of di-*t*-butyl disulfide. For this disulfide the maximum absorbance occurs at a shorter wavelength than for normal disulfides. In this molecule, it might be expected that the very bulky *t*-butyl groups, through van der Waals repulsion, would tend to increase the dihedral angle in order to increase their separation. It thus appears that decreasing the preferred 90° dihedral angle of a disulfide, as by putting it in a five-membered ring, shifts absorbance to a longer wavelength (but with decreased intensity) while increasing the preferred 90° dihedral angle, as in di-*t*-butyl disulfide, has the opposite effects, shifting absorbance to a shorter wavelength (but with increased intensity).

MELTING POINTS OF MOLECULAR CRYSTALS

The process of transforming the orderly array of molecules in a crystal to the disorderly array in the melt involves two of the properties of molecules we have been discussing, the preferred geometry of the molecule and the strength of intermolecular forces. At the melting point T_m the process of interconversion of crystal and liquid is at equilibrium. By definition, the free-energy change ΔG at equilibrium is zero. Therefore, for $T_m = $ mp, the enthalpy

$$\Delta G_{fus} = \Delta H_{fus} - T_m \Delta S_{fus} = 0$$

or

$$\Delta H_{fus} = T_m \Delta S_{fus}$$

of melting ΔH_{fus} favoring the crystalline state is exactly counterbalanced by the entropy of melting ΔS_{fus} favoring the liquid state. The enthalpy for the crystal is lower, i.e., the molecules are at a lower potential energy, than for the liquid, so that this positive enthalpy of fusion will always favor the crystalline state, i.e., tend to make ΔG_{fus} a positive number. Since the liquid state is more disordered, the entropy of the liquid will be greater than that of the crystal; that is, ΔS_{fus} is also positive. A positive value for ΔS_{fus}, however, will tend to make ΔG_{fus} *negative*, especially as the temperature increases. The melting point is thus the temperature at which the enthalpy term ΔH_{fus} favoring the crystal is just balanced by the entropy term $-T\Delta S_{fus}$ favoring the liquid.

Table 3-5 summarizes data on ΔH_{fus} and ΔH_{vap} for a number of simple molecules, together with their melting and boiling points. It can be seen that there is a relatively close parallel between the order of the compounds by increasing ΔH_{vap} and by increasing boiling point. This is not so accurately true for the melting points, and in almost every case the discrepancy can be ascribed to differences in molecular symmetry.

In general, the more unsymmetrical a molecule, the greater the change in entropy in going from the liquid to the crystal state will be. For molecules with similar cohesive energies, this may be illustrated by argon, a spherically symmetrical molecule, compared to nitrogen and oxygen, with axial symmetry. Similarly, the symmetrical molecule carbon tetrachloride melts higher than the less symmetrical molecules of chloroform and hydrogen fluoride. Another example from Table 3.5 is neopen-

Table 3.5 Enthalpies of Fusion and of Vaporization

	ΔH_{fus}, $kcal$ $mole^{-1}$	ΔH_{vap}, $kcal$ $mole^{-1}$	mp, $°C$	bp, $°C$
He	0.005	0.020	< -272.2	-269
H_2	0.028	0.216	-259	-253
N_2	0.172	1.33	-210	-196
Ar	0.265	1.56	-189	-186
O_2	0.106	1.63	-218.4	-183
CH_4	0.225	1.95	-183	-161
CO_2	1.98	3.52	-57	-78
$C(CH_3)_4$	*	5.4	-17	9.5
$n\text{-}C_5H_{12}$	*	6.3	-130	36
H_2O	1.44	10.52	0	100
HF	1.10	7.2	-83	20
HCl	0.48	3.9	-111	-85
H_2S	0.57	4.5	-83	-60
NH_3	1.35	4.5	-78	-33
PH_3	0.27	3.5	-132.5	-85
CCl_4	0.64	8.0	-23	77
$CHCl_3$	*	7.3	-63.5	61

*Note that the ΔH_{fus} is generally about one-tenth the ΔH_{vap}.

tane, with great symmetry and very little flexibility, and *n*-pentane, with much less symmetry and the flexibility arising from conformers at the two central carbon-carbon bonds. In order for *n*-pentane to fit into a crystal structure, both these bonds must be in the trans conformation and the axis of the trans-trans conformers must be properly oriented with respect to the molecules which will be its neighbors in the crystal.

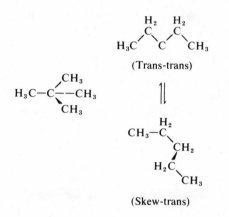

(Trans-trans)

(Skew-trans)

Another example which might be cited to illustrate the importance of a symmetrical fixed conformation is the marked difference in melting point for C_2Cl_6 (187°) compared to Si_2Cl_6 (−1°). As is indicated in Table 3.2, C_2Cl_6 has a very large barrier to rotation and thus would be a "rigid" molecule while Si_2Cl_6 has an extremely low barrier to rotation and would rotate much more freely at the Si—Si bond. Thus *n*-pentane and Si_2Cl_6 illustrate the influence of conformational freedom in lowering melting points, as compared to neopentane and C_2Cl_6, respectively.

In molecules with rigid geometry, such as cis-trans isomers, there is a marked factor making the more symmetrical isomer the higher-melting, even though it may have less cohesive energy. For example, *cis*-1,2-dichloroethylene has a much lower melting point than the more symmetrical trans isomer. The same tendency is shown for the *cis-* and *trans*-2-butenes.

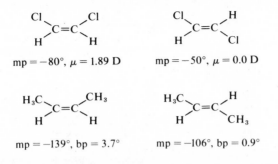

mp = −80°, μ = 1.89 D   mp = −50°, μ = 0.0 D

mp = −139°, bp = 3.7°   mp = −106°, bp = 0.9°

This influence of molecular symmetry is also shown in the melting points of the isomers anthracene and phenanthrene and of the dichloronaphthalenes, where the 2,6 isomer, the most symmetrical, is the highest-melting and the 1,2 isomer, the least symmetrical, is the only one melting below room temperature.

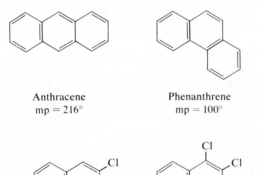

Anthracene
mp = 216°

Phenanthrene
mp = 100°

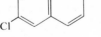

2,6-Dichloronaphthalene
mp = 135°

1,2-Dichloronaphthalene
mp >25°

The introduction of an oxygen atom between the two phenyl groups of the axially symmetrical biphenyl molecule considerably lowers its melting point and, of course, sharply bends the molecule.

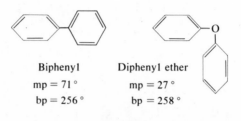

Biphenyl	Diphenyl ether
mp = 71°	mp = 27°
bp = 256°	bp = 258°

PROBLEMS

3.1 From the ratio of the conformers given for 1,1-difluoro-1,2-dibromo-2,2-dichloroethane, calculate the energy difference ΔF for the trans and skew forms, using Eq. (4).

3.2 Write an equation like Eq. (4) for the axial-equatorial conformers. Use it to estimate the percent of axial and equatorial conformers in equilibrium at 300 K for methylcyclohexane and for *t*-butylcyclohexane.

SUGGESTED READING

Eliel, E. L.: "Stereochemistry of Carbon Compounds," chaps. 5 – 8, McGraw-Hill Book Co., New York, 1962.

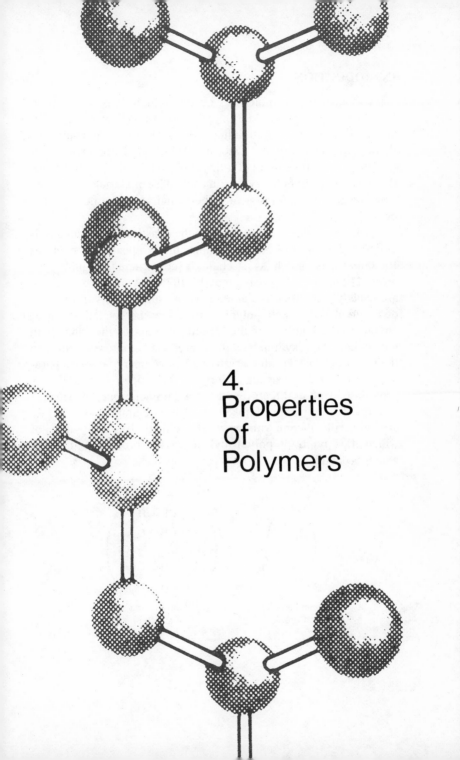

4.
Properties
of
Polymers

INTRODUCTION

Very long chainlike molecules occur in such extremely important natural products as starch, cellulose, rubber, protein, RNA, and DNA, and are the building blocks of such important synthetic materials as nylon, Dacron, Orlon, butyl rubber, silicone rubber, polystyrene and poly(vinyl chloride). It is the purpose of this chapter to explore why some of these materials are moldable plastics, some form strong fibers, and some are flexible rubbers.

The chainlike molecules which constitute the materials mentioned above are built of simpler units repeated as many as 100,000 times, much as a chain is constructed of individual links. These polymer chains may be 10,000 Å or more long and are usually 5 to 10 Å in diameter. How can the gross properties of assemblies of such polymer molecules be related to three characteristic features of the structural units of the long-chain molecules: their symmetry, their preferred conformation, and their van der Waals interactions (cohesive-energy-density parameter)? In other words, can we relate the properties of a fibrous protein to structural features of the amino acid units of which it is built, of starch and cellulose to the glucose units of which they are built, of silicone rubber to the dimethylsiloxane units of which it is built, of polypropylene to the propylene units of which it is built?

Amino acid unit
of protein

Glucose unit
of starch or
cellulose

Dimethylsiloxane
unit of silicone
rubber

Propylene unit
of polypropylene

The overall shape of long-chain polymer molecules can be (1) rodlike, e.g., polyethylene, nylon, cellulose (when in crystalline form), (2) helical, e.g., isotactic polypropylene in the crystalline state, DNA, and some fibrous proteins, or (3) randomly coiled [all rubbers including atactic polypropylene, polystyrene, poly(methyl methacrylate), as well as most polymers in solution.] Let us now see how such characteristics of the units of the chain as conformation, symmetry, and cohesive energy density are typically reflected in the properties of polymeric materials.

POLYETHYLENE

The simplest polymer chain is polyethylene, prepared by polymerization of ethylene.

$$n\ CH_2{=}CH_2 \xrightarrow{\text{cat}} {+}CH_2CH_2{+}_n$$

Ethylene   Polyethylene, mp 140°

$$2n\ CH_2N_2 \xrightarrow{\text{cat}} \text{Polyethylene} + 2nN_2$$

Diazomethane

The same polymer can be prepared in the laboratory from diazomethane. The preparation from ethylene is a major commercial operation. The product is used as a moldable plastic, as a film, and in the form of fibers in rope, hawsers, and rugs.

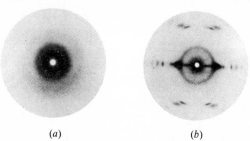

(a)   (b)

Fig. 4.1 *(a) X-ray diffraction pattern for quenched polyethylene, and (b) a stretched polyethylene fiber. (From C. E. H. Bawn, "The Chemistry of High Polymers," following p.* 190, *Interscience Publishers, Inc., New York,* 1948.)

A study of the x-ray diffraction patterns for polyethylene has shown it to be a highly crystalline material. Detailed analysis of such patterns reveals the exact array of atoms and molecules in the polyethylene crystal. The C—C bond length is 1.54 Å, the C—H bond length 1.08 Å, all bond angles are tetrahedral (109.5°), and the carbon atoms are in the extended trans conformation.

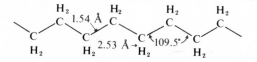

The distance of 2.53 Å is referred to as the *repeat distance* along the chain axis, i.e., the distance along the axis of the chain in the crystal between groups of identical orientation. In other words, if we displaced one ethylene unit by 2.53 Å along the chain axis, it would exactly correspond in space to the adjacent unit.

Since butane prefers the trans over the two skew conformations by about 0.9 kcal mole^{-1}, that is, 72 percent of the molecules at room temperature are in the trans form (Chap. 3), the conformation in crystalline polyethylene represents the energetically preferred arrangement of the chain.

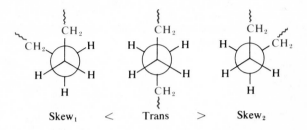

Skew₁   <   Trans   >   Skew₂

Since even in the melt or in solution a majority of the conformations will already be in this necessary arrangement for the crystal, the remaining reorganization required to get all bonds into the necessary trans arrangement is more than provided for by the heat of crystallization. The chains thus crystallize readily and display a melting point of 140°.

There is a lower-melting, somewhat softer form of polyethylene, made by an earlier high-temperature, high-pressure process. Its molecular structure has been shown to deviate from the product of the low-temperature catalytic process we have been discussing in that its molecules have a number of *branches* in the

chain. In the immediate vicinity of such a branch, the three chain segments cannot fit into the normal crystal domains of polyethylene. These regions thus remain uncrystallized, amorphous, and rubbery. As a result the crystallites are smaller and 20 to 40 percent of the material remains amorphous and rubbery, giving a softer, more flexible product.

Table 4.1 Change in Melting Point with Molecular Weight for Linear Polyethylene

M.p., °C	Av. mol. wt.	M.p., °C	Av. mol. wt.
140.7	200,000	114	1,400
138	14,000	105	980
135	7,000	99	840
126	2,800	81	560

In addition to chain branches serving as an "imperfection" which will not fit into the crystal structure, chain ends can also be considered as imperfections. Thus, decreasing the average molecular weight of polyethylene will increase the concentration of chain-end imperfections, and this will consequently lower the melting point of the polymer. Data on the variation of melting point with molecular weight are summarized in Table 4.1.

This dependence of melting point on chain length has also been investigated for poly(ethylene oxide) and crystalline (isotactic) poly(propylene oxide); some results are given in Tables 4.2 and 4.3.

$$n \; CH_2{-}CH_2 \quad \xrightarrow{\text{cat}} \quad {\left(OCH_2CH_2\right)}_n$$
$$\diagdown O \diagup$$

Ethylene oxide

$$n \; CH_3CH{-}CH_2 \quad \xrightarrow{\text{cat}} \quad {\left(\overset{\displaystyle CH_3}{\underset{\displaystyle |}{OCHCH_2}}\right)}_n$$
$$\diagdown O \diagup$$

Propylene oxide

These data all indicate that shorter chains have proportionally lower melting points. In fact, as an approximation, each chain end can be considered as an impurity so far as the crystal structure is concerned. Since the equation relating freezing-point depression ΔT to molecular weight is $\Delta T = K_f m$, where m is the molal concentration of impurities, i.e., moles per 1,000 g of

Table 4.2 Change in Melting Point with Molecular Weight for Poly(Ethylene Oxide)

M.p., °C	*Av. mol. wt.*
65	200,000
58	3,270
43	1,090
22	536
11	424

solute, and K_f is a constant characteristic of the crystallizing matrix, values for K_f for polyethylene, poly(ethylene oxide) and poly(propylene oxide) can be calculated from the data in Tables 4.1 to 4.3 (see Prob. 4.1).

Table 4.3 Change in Melting Point with Molecular Weight for Isotactic Poly(Propylene Oxide)

M.p., °C	*Av. mol. wt.*
75	200,000
64.5	18,000
56	1,940
36	920

POLYISOBUTYLENE

The properties of polyisobutylene are quite different from polyethylene, even though it is also a saturated hydrocarbon polymer.

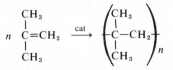

Isobutylene   Polyisobutylene

Polyisobutylene is a rubbery material (the basic building block of butyl rubber), and when unstressed, it shows no crystallinity under examination by x-ray diffraction. It will solidify to a glassy, *noncrystalline* amorphous solid at temperatures below $-70°$. This temperature at which a rubber or viscous polymer solidifies to an amorphous glass is referred to as the *glass transition temperature* T_g. This transition from solid glass to liquid (or rubber) is not as sharply defined as the melting point of a crystal T_m.

Why should these hydrocarbon polymers be so different, with one having $T_m = 140°C$, the other $T_g = -70°C$? At least one major reason can be seen from an examination of the conformers of the units involved.

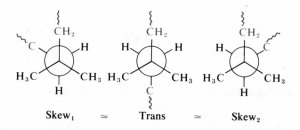

Skew$_1$ $\simeq$ Trans $\simeq$ Skew$_2$

In this case, the polymer chain on the rear carbon (C∿) has the choice of being between two methyl groups, CH_3, in the trans conformer, or between a methyl group and the polymer chain, CH_2∿. Since the methyl group, CH_3, and methylene group, CH_2∿, have very similar steric requirements, there will be little or no difference in energy between these conformations. In the absence of any significant energy differences, entropy will dictate the most random arrangement, i.e., roughly equal populations of skew$_1$ (~33 percent), trans (~33 percent), and skew$_2$ (~33 percent). Such a population of conformers would make the preferred arrangement for the polyisobutylene chain (on entropy grounds) a random coil. Since the energy barrier for interconversion of these conformers is low enough for interconversion to occur 10^4 times per second or more at room temperature, the chain will not only be randomly coiled but will be highly flexible in responding to stresses exerted for longer than a hundredth of a second or so.

These are indeed the requirements for a polymer molecule to exhibit rubbery properties. The equilibrium conformation is a random coil. On stretching, this can be extended to a more linear conformation. On release, it will rapidly return to the more stable random coil. For an ideal rubber, the free-energy change on stretching can be shown to be entirely a change in entropy; i.e., there is no enthalpy change between the unstretched and stretched state or

$$\Delta G = -T \ \Delta S$$

For polyisobutylene, for example, this is consistent with the view that the skew and trans forms have the same enthalpy. However, on extending a chain from random coil one must increase the relative number of trans conformations at the expense of skew. One is thus proceeding from the disordered random coil (high entropy) to the ordered extended chain (low entropy). The random coil thus has a lower free energy than the extended chain; the extended chain will thus spontaneously revert to the random coil.

Stretching a rubber band is thus a process for which the entire change in free energy (increasing with stretch or decreasing on relaxation) is associated with a change in entropy between the disordered random coils in relaxed rubber and the more ordered extended chain in stretched rubber. The change in entropy on stretching rubber has an important consequence on the tendency of rubber to crystallize, i.e., on its melting point. Recall that the melting point, the equilibrium between the crystalline and liquid states, is governed by the relation

$$\Delta H_{fus} = T_m \ \Delta S_{fus} \tag{1}$$

It is important, incidentally, to realize that the rubbery state is indeed a special example of the liquid state. It is a liquid in which the very long-chain polymer molecules permit deformation by changing the shape of the polymer chain, including motion of units in one chain not only with respect to the same chain but with respect to units in neighboring chains. However, overall "flow" of the entire molecule is retarded either by the major entanglement of the long-chain molecules or by their occasional *cross-linking* into a three-dimensional network structure, e.g., by vulcanization with sulfur or by the existence of very small crystallites which serve as network points.

From the thermodynamic point of view, then, the rubbery state is a liquid state, and the equilibrium process we are to consider is between this disordered liquid state of random coiled molecules and an orderly array of the polymer chains in a crystal. The ΔH_{fus} will remain essentially unaffected by the stretching of the rubber, but, as we have pointed out above, stretching will markedly affect the entropy of the flexible (or liquid) state of rubber. The decrease in entropy of stretched rubber will diminish the entropy change on going from the liquid to the crystalline state. In other words, stretching rubber orients the polymer chains and facilitates their crystallization. Thus, since ΔS_{fus} decreases with stretch while ΔH_{fus} is essentially constant, Eq. (1) requires that T_m *increase* with stretching.

For example, an unstretched butyl rubber band shows no evidence of crystal order by x-ray at room temperature. On cooling below its glass transition temperature, it becomes hard and inflexible but still shows no evidence of crystal order by x-ray; i.e., it is an amorphous solid. On stretching such a rubber band to 4 or 5 times its original length, however, it develops a characteristic crystalline x-ray diffraction pattern (Fig. 4.2). In other words, by stretching butyl rubber sufficiently, we have so lowered the unfavorable entropy change between liquid and crystal that the material "crystallizes" at room temperature. Stretching has raised the melting point T_m above room temperature.

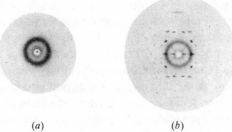

(a)   (b)

Fig. 4.2 *X-ray diffraction pattern for (a) un-stretched (amorphous) polyisobutylene and for (b) sample stretched 1,400 percent. (From C. E. H. Bawn, "The Chemistry of High Polymers," following p. 190, Interscience Publishers, Inc., New York, 1948.)*

Detailed study of the crystal x-ray pattern for butyl rubber has revealed one other feature of significance to the geometry of polymer molecules. The most extended form of a carbon chain (without deforming the tetrahedral bond angles) is the all-trans conformation illustrated by the polyethylene chain. In crystalline polyisobutylene (butyl rubber), however, it turns out that the chain cannot assume this completely extended linear zigzag conformation, because of methyl-methyl repulsions.

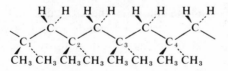

Extended, all-trans polyisobutylene

If one makes space-filling models of this chain, it readily becomes apparent that the van der Waals radii of the methyl groups overlap significantly. The hydrogens on these methyls are separated only by 1.85 Å, rather than the sum of van der Waals radii for hydrogen of 2.4 Å. This strain is relieved by systematically distorting the ideal all-trans conformation so that the methyl groups on C_2 are not directly opposed to those on C_1. The actual conformation proposed is a rather complex helix involving eight monomer units in the repeat distance of 18.6 Å. Note that this distance is slightly less than that for eight units in polyethylene (20.16 Å) since the polyisobutylene chain is not as fully extended as the all-trans chain of polyethylene.

Another mode of accommodating steric hindrance can be seen in the crystal structure of poly(vinylidene chloride), an important plastic known as Saran.

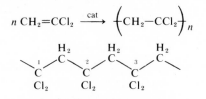

All-trans poly(vinylidene chloride)

In this chain, the fully extended all-trans form would have interference between the chlorine atoms, comparable to that for the methyl groups in polyisobutylene. [See Table 3-2 for a comparison of the barrier for chlorine in ethyl chloride (3.56 kcal mole^{-1}) with methyl in propane (3.30 kcal mole^{-1}).] For crystalline poly(vinylidene chloride), the x-ray data indicate a repeat distance of 4.7 Å, which corresponds to only two vinylidene chloride units and indicates a systematic folding in the chain conformation and a consequent shortening over 5.06 Å that would be observed in the fully extended chain. The chlorine atoms on C_2 have rotated 45° to decrease chlorine-chlorine repulsions, but the chlorines on C_3 rotate back 45° rather than further in the same direction. This gives C_3 an identical orientation to C_1 and accounts for the repeat distance of 4.7 Å. Note that 4 times 4.7 Å (18.8Å) is very close to the repeat distance observed for eight units in polyisobutylene.

NATURAL RUBBER AND BALATA

These two natural polymers of isoprene have been shown to have different geometry at the carbon-carbon double bond. In natural rubber, the polymer chain is cis at the double bond; in balata it is trans.

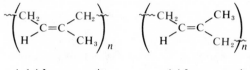

cis-1,4-Isoprene unit
in rubber

trans-1,4-Isoprene unit
in balata

Natural rubber does not readily crystallize at room temperature even though it has a crystal melting point above 20°C and normally remains rubbery down to its glass transition temperature T_g of −70°C, whereas balata is a hard crystalline polymer at room temperature and has been used for such purposes as golf ball covers. This difference in properties is consistent with the more extended and symmetrical structure for the trans-1,4-isoprene unit as compared to the bent and less symmetrical structure for the cis unit (compare these properties for rubber and balata to the melting points for *cis*- and *trans*-2-butene).

GLASSY POLYMERS

The preceding discussion has emphasized two essential characteristics of a linear polymer necessary for the display of rubbery properties: (1) a preference for a randomly coiled chain conformation and (2) that the chains move easily and rapidly under stress. The latter requirement demands low enough barriers to chain rotation and to interchain segment motion (low CED) to permit rapid uncoiling of chains under stress and rapid recoiling on relaxation. A polymer illustrating this point is poly(methyl methacrylate) (Plexiglas or Lucite).

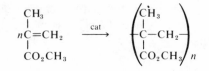

Methyl methacrylate   Poly(methyl methacrylate)

The geometrical features of this polymer chain resemble polyisobutylene insofar as there are two bulky groups attached to every other carbon of the polymer chain. In contrast to polyisobutylene, which has a glass transition temperature of $-70°$, poly(methyl methacrylate) is an amorphous solid with a T_g of 70°C. Thus the chains of both polyisobutylene and poly(methyl methacrylate) prefer a randomly coiled arrangement, but for polyisobutylene these chains are not "immobilized" until temperatures below $-70°C$ are reached while for poly(methyl methacrylate) they are immobilized at temperatures below 70°C.

One reason for this greater restriction to mobility in poly(methyl methacrylate) is the greater van der Waals attraction (higher cohesive energy density) between chain units. In polyisobutylene, there are only nonpolar C—C bonds and very weakly polar C—H bonds. In poly(methyl methacrylate), we have introduced much more polar C—O and C=O bonds. These oxygen atoms have thus introduced the possibility of enhanced van der Waals attractions through these dipoles. In poly(methyl methacrylate), this increased attraction between chains means that only at temperatures above 70° is the thermal kinetic energy $(3RT/2)$ great enough to allow segmental motion, thus permitting the flow necessary for rubbery or liquid behavior.

The marked difference in properties between polyisobutylene and poly(vinylidene chloride), $T_g = -17°C$, can also be ascribed to the difference in CED, since the latter polymer contains polar C—Cl bonds introducing additional atttractive forces between chain units.

Table 4.4 Values for CED Parameter and for Glass Transition Temperature for Various Polymers

Polymer	CED, cal ml^{-1} *	T_g, °C
Silicone rubber	~50	−125
Ethylene	62	−70
Isobutylene	65	−70
Natural rubber	67	−70
Styrene	74	80
Methyl methacrylate	83	70
Vinyl acetate	88	28
Ethylene terephthalate (Dacron)	114	90
Hexamethylene adipamide (nylon)	185	−10
Acrylonitrile	237	

*Compare these values for polymers with those for low-molecular-weight liquids in Table 2.4.

Note in Table 4-4 that there is no accurate parallel between CED and T_g. This is because an increase in T_g may be caused not only by greater cohesive forces between units in adjacent chains but by factors decreasing the flexibility of chains. For example, the bulky phenyl group, C_6H_5, attached to every other atom in the polystyrene chain greatly limits the conformational freedom and thus the flexibility of the chain, which is reflected in an unexpectedly high T_g.

Polymers such as poly(methyl methacrylate) can exhibit rubbery properties at temperatures above the glass transition temperature. This accounts for a phenomenon known as *plastic memory*. If, for example, a straight rod of poly(methyl methacrylate) is heated above its T_g (say 125°C), bent, and cooled rapidly, it will congeal to the glassy state in the bent shape. Upon heating at any later time, however, it will revert to its original shape. This is much as though we stretched a rubber band and then cooled it in Dry Ice or liquid nitrogen. It would congeal to a solid in the stretched form. If the rubber band warmed up again above the glass transition temperature, the molecules which were immobilized in a stretched conformation at low temperature would acquire sufficient thermal energy to become mobile and then prefer to return to their random-coil arrangement. The rubber band which had been immobilized in the stretched shape at very low temperature, would then be able to relax to its original shape on warming. Exactly the same considerations apply to poly(methyl methacrylate), except that the transition from glass to rubber occurs above 70°C, and so poly(methyl methacrylate) is glass at room temperature and becomes rubbery only at elevated temperature, whereas polyisobutylene is rubbery at room temperature and becomes a glass only at very low temperatures. Thus the bent rod of poly(methyl methacrylate), on warming above the glass transition temperature, will become rubbery and will revert to its original unstressed straight form.

Perhaps a useful analogy between the glassy and rubbery states is a bowl of cooked spaghetti. When hot, the sauce is fluid and readily permits one strand to slip past another; the assembly of intertwined spaghetti is fluid and will assume the shape of any container in which it is placed. If the spaghetti is chilled, the sauce hardens, the chains can no longer flow readily, and the bowl of spaghetti "solidifies." If it is turned out, it will retain the shape of the bowl in which it congealed. Like all analogies, there are dangers in this one. Actually, the long-chain polymers are not "dead," like a piece of spaghetti, but are "alive" with thermal kinetic energy, so that a can of worms might be a somewhat better analogy. Furthermore, the hardening of an assembly of polymer molecules is not from a change in the sauce, i.e., the cohesive energy between chains, but a decrease in the liveliness of the chains themselves at lower temperature.

VULCANIZATION AND VISCOUS FLOW

It is important to distinguish between the process of elastic deformation of a rubber and viscous flow. For poly(methyl methacrylate), for example, or for Silly Putty both processes can occur under the proper circumstances. To anyone who has not had the pleasure of playing with Silly Putty (an unvulcanized silicone rubber), we should explain that it is a material of puttylike consistency which, when molded in the form of a ball, will bounce vigorously. The properties can be explained on the basis that it is a viscous liquid made of rubbery polymer chains.

When Silly Putty is deformed rapidly, the deformation is accommodated by deforming silicone rubber molecules from their preferred random-coil shape. If the deformation is rapidly released, the molecules rapidly return to their original conformation. This is what happens in the fraction of a second the bouncing ball is in contact with the floor. If the deforming force is maintained for a longer time (seconds or minutes), the deformed molecules can slowly resume an equilibrium random-coil conformation by viscous flow, by which the molecules assume a new equilibrium position with respect to other molecules.

Vulcanization prevents this process of cold flow under stress by cross-linking the polymer chains by covalent bonds. For example, when a diene rubber, natural or synthetic, is heated with sulfur, chemical links between chains are established.

$$\sim(CH_2CH{=}CHCH_2)\sim \qquad \sim CH_2CH{=}CH{-}CH\sim$$
$$\xrightarrow[\text{heat}]{S} \qquad\qquad\qquad \overset{|}{\underset{|}{S}}$$
$$\sim(CH_2CH{=}CHCH_2)\sim \qquad \sim CH_2CH_2{-}CH{-}CH_2\sim$$

1,4-Butadiene units
in rubber

A sulfur atom thus ties two rubber chains together. If this process occurs more than twice for each original polymer molecule, the entire assembly of rubber molecules becomes united into one gigantic three-dimensional molecule. It is then no longer possible for viscous flow to occur under stress, so that *vulcanized* rubber always returns to its original shape when the stress is relaxed.

COLD-DRAWING OF FIBERS

A process which bears some mechanical relationship to stretching a rubber band is that of cold-drawing spun fibers, either natural or synthetic. The spider and the silkworm do this for their protein fibers just as industry does for nylon or Dacron. In industry the original polymer fiber is spun by forcing molten polymer through a fine orifice. As the fine thread of viscous melt cools, it crystallizes, at least partially.

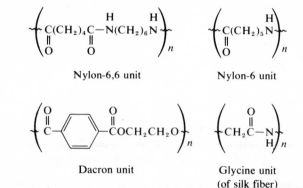

Nylon-6,6 unit   Nylon-6 unit

Dacron unit   Glycine unit
(of silk fiber)

These long-chain polymers useful for fibers have relatively high crystalline melting points. The two nylon types have long chains resembling polyethylene but with the amide group, $-CONH-$, built in at regular intervals. This group is also a characteristic feature of protein molecules, as illustrated above by the glycine unit of silk fiber. It contributes particularly strong van der Waals bonding forces between chains.

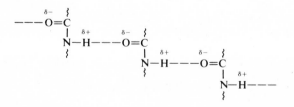

The intermolecular electrostatic attraction involving a positive charge on hydrogen with a negative oxygen (or nitrogen) is an especially strong force because of the shorter van der Waals radius for hydrogen. The force is further enhanced for amides because the bond moment at a $\diagdown \text{C}=\text{O}$ group puts a much larger charge on the oxygen (-1.93×10^{-10} esu) than for $\text{C}-\text{O}$ (-0.60×10^{-10} esu; see Table 2.2). Thus the so-called *hydrogen bonding* illustrated above and in Fig. 4.3 for amides makes a very large contribution to the CED, giving nylon-6, 6, nylon-6, and silk fibers very much higher melting points than polyethylene.

For Dacron, the high melting point can be ascribed in part to a modest increase in the CED through the introduction of polar $\diagdown \text{C}=\text{O}$ groups and also to the symmetry and rigidity of the para-substituted benzene ring. The flat planar and linear structure of this part of the polymer unit makes the entropy difference between melt and crystal less and thus favors a higher melting point.

On cooling, the disordered array of polymer molecules in nylon or Dacron tend to crystallize. The cooled fibers do indeed show a crystal pattern by x-ray diffraction, but the pattern resembles that of a crystalline powder. That is, the individual small crystals are in random orientation, so that instead of a pattern of diffraction spots it is one of diffraction rings. On stretching such a cooled spun fiber, an interesting and remarkable phenomenon occurs. A section of the fiber suddenly "necks down" to about one-half or one-third the original diameter, and as stretching continues, the segment of necked-down fiber grows at the expense of the original. When the entire fiber has necked down, the resistance to further stretching suddenly increases sharply. The strength of this cold-drawn fiber is markedly greater than the original. If a new x-ray pattern is made, it now resembles that for a single crystal, with diffraction spots (see, for example, Fig. 4.1).

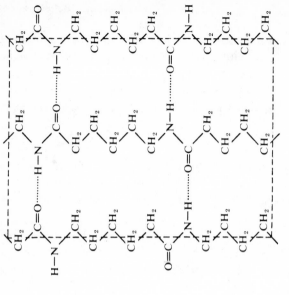

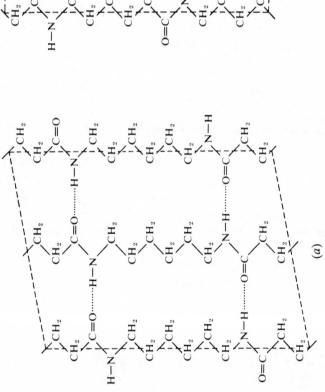

Fig. 4.3 Arrangement of chains in hydrogen-bonded sheets in the crystal structure of (a) nylon-6,6 and (b) nylon-6. (Redrawn from Fred W. Billmeyer, Jr., "Textbook of Polymer Science," Interscience Publishers, a Division of John Wiley & Sons, Inc., New York, 1966.)

The interpretation of this phenomenon is that the crystal structure in the undrawn fiber consists of small crystal domains (or crystallites) more or less randomly oriented with respect to the fiber axis. On cold-drawing, two processes occur: (1) the unoriented, amorphous domains become oriented along the fiber axis by stretching and tend to become more crystalline, and (2) the original crystallites which were not oriented along the fiber axis become reoriented in this way. Figure 4.4 represents these concepts. The cold-drawing process thus increases the fraction of material in crystal domains (at the expense of amorphous) and orients the crystal domains with the polymer-chain axis coincident with the fiber axis. This process differs from the similar orientation of polymer chains on stretching a rubber since the strong crystal forces in nylon and Dacron hold the chains in the new conformation.

Crystal domains
(crystallites)

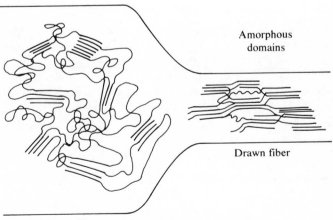

Amorphous
domains

Drawn fiber

Undrawn fiber

Fig. 4.4 *Diagrammatic sketch of changes made by cold-drawing fibers.*

POLYPROPYLENE AND STEREOREGULAR POLYMERS

So far we have discussed the influence of chain conformation and CED on the properties of linear polymers, as well as some aspects of chain length, chain branching, and chain cross-links. One other factor of major significance to both natural and synthetic properties is the configuration of chain units when they are chiral. A simple example is polypropylene. Depending on the catalyst system used for its preparation, polypropylene can be a rubbery polymer with $T_g = -25°$ or a crystalline fiber-forming polymer with $T_m = 170°$.

The difference in x-ray diffraction patterns (Fig. 4.5) indicates the difference in order involved.

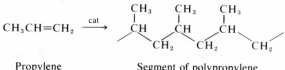

Propylene   Segment of polypropylene

The nature of the geometric or stereochemical difference possible for polypropylene can be seen by examining the segment above. The placement of each methyl group in this segment could be on the front or the rear side of the plane represented by the carbon atoms of the polymer-chain backbone in its trans conformation. Each methyl group could have the same *configuration*; i.e., all methyl groups on the same side. This arrangement of chiral groups in the same configuration has been designated as *isotactic*.

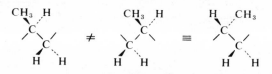

Chiral forms (or enantiomorphs)
(nonsuperimposable mirror images)

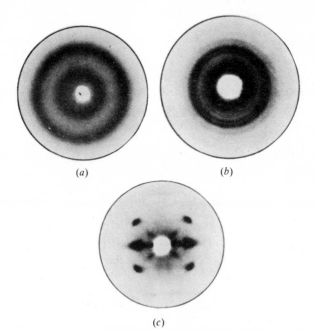

Fig. 4.5 *X-ray diffraction diagrams of polypropylene: (a) atactic polymer; (b) unoriented isotactic polymer; (c) drawn and oriented isotactic fiber. (From N. G. Gaylord and H. F. Mark, "Linear and Stereoregular Addition Polymers," p. 54, Interscience Publishers, Inc., New York, 1959.)*

The methyl groups could have alternate configurations, i.e., one in front and each adjacent one in back. This regular alternating configuration of the chiral groups is called *syndiotactic*. There is also a random arrangement of configurations which is called *atactic*. The isotactic sequence is one where the probability of each adjacent group's having the same configuration is *unity*, syndiotactic where this probability is *zero*, and atactic where this probability is *one-half*. One can thus envision any degree of *tacticity* from 100 percent isotactic to 100 percent syndiotactic

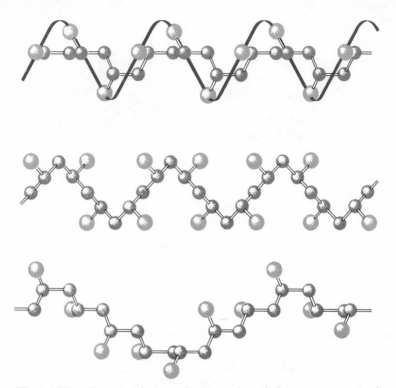

Fig. 4.6 *Three-dimensional views of polypropylene (hydrogen atoms omitted). Colored balls represent methyl groups,* CH_3. *In isotactic polypropylene (top) the* CH_3 *groups define a helix (color ribbon). In syndiotactic polypropylene (middle) structure is also regular. Atactic polypropylene (bottom) is nonregular in form. (Redrawn from Giulio Natta, Precisely Constructed Polymers. Copyright © 1961 by Scientific American, Inc. All rights reserved.)*

with atactic being the special case of 50:50. Note that the *configuration* (isotactic, syndiotactic, or atactic) is built into the polymer and is not changed by altering the *conformation*. The three configurations above will represent different structures whether we represent the chain in an all-trans conformation (as above) or all-skew$_1$ or all-skew$_2$.

In the case of polypropylene, the crystalline form (melting point 170°) has been shown by x-ray diffraction to be isotactic, whereas the amorphous rubbery form is the atactic configuration. Syndiotactic polypropylene has also been prepared. It is crystalline with a repeat distance of 7.4 Å involving four monomer units. Schematic representations of the carbon atoms in isotactic and syndiotactic polypropylene are shown in Fig. 4.6.

All three forms are now known for several other chiral polymers such as poly(methyl methacrylate), poly(1,2-butadiene) and poly(*t*-butylethylene oxide).

1,2-Butadiene unit   *t*-Butylethylene oxide unit

The x-ray diffraction evidence for isotactic polypropylene fibers has demonstrated that this polymer does not crystallize in an all-trans conformation of the backbone but in a regular helical conformation. The reasons for this regular helical-chain conformation can be recognized by examining the possible conformations. For this molecule, if we have a configuration with

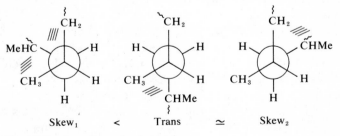

Fig. 4.7 *Conformations of polypropylene.*

the methyl group on the front carbon on the lower left, then the skew$_1$ form will put the bulky chain in the rear between two bulky groups, while skew$_2$ will put the chain in the rear next to the chain methylene group, $-CH_2\sim$, very similar to its environment in the trans conformation next to a methyl, $-CH_3$. The diagrams thus lead to the proposition that trans and skew$_2$ will be about equal in energy while skew$_1$ will be definitely unfavorable. Thus if all the methyl groups have the same configuration, the chain will tend to skew in just one sense, as is necessary to form a regular helix. The other important hindrance factor which helps to shape the preferred conformation of isotactic polypropylene in the crystal is the overlap which would occur for adjacent methyl groups in the all-trans conformation. As was pointed out in the discussion of the stretched orientation for polyisobutylene, the all-trans chain conformation leads to an interpenetration of van der Waals radii of opposing methyl groups on alternate carbon atoms. Two consecutive trans conformations in isotactic polypropylene are thus disfavored by this van der Waals repulsion between adjacent methyls, which is entirely relieved by alternate trans, skew$_2$, trans, skew$_2$ conformations.

This conformation does provide a regular helical arrangement for the chains in the crystal, in agreement with the repeat distance observed by Natta of 6.7 Å. Each skew$_2$ bond introduces a twist of 120°, so that three are required to produce a monomer unit in the same orientation as the first. Figure 4.6 illustrates the helical conformation of crystalline isotactic polypropylene, with a repeat distance involving three propylene units.

For atactic polypropylene, in which the configuration of the methyl group could be left as well as right (see Fig. 4.6), the preferred skew could be right as well as left. This would, of course, produce a randomly coiled molecule, as is necessary for the disordered rubbery state of this polymer.

For isotactic poly(propylene oxide), the crystal has been shown to have a backbone in the all-trans conformation.

Propylene oxide

Atactic, rubber, $T_g \sim -60°$
Isotactic, crystal, $T_m = 75°$

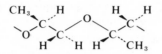

All-trans conformation for isotactic poly(propylene oxide)

The extra spacing with an oxygen between each propylene unit eliminates the methyl overlap on alternate carbons which ruled against the all-trans conformation for isotactic polypropylene, so that for isotactic poly(propylene oxide) x-ray diffraction data show that it does indeed have the all-trans conformation along the polymer backbone.

Incidentally, the lower T_g for atactic poly(propylene oxide) ($-60°$) than for atactic polypropylene ($-25°$), despite the greater van der Waals forces which would be introduced by the polar C—O bonds in the former, can be rationalized in terms of the significantly lower barrier to rotation at the C—O bond (see Table 3-2). The oxygen will occupy less space than a CH_2 group (see the comparison of CH_2 and O in the cyclohexane ring system, Chap. 3), and this factor will also confer greater conformational freedom upon the oxide polymer. It is this greater conformational mobility which keeps the chains flexible enough to be rubbery down to $-60°$, a feature which is still further accentuated in the silicone rubbers.

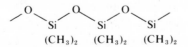

Dimethylsiloxane units of silicone rubber

Since the barriers to rotation at the silicone-oxygen bond are still lower than for C—O bonds, it takes an even lower temperature to "freeze out" rotation between conformers. The T_g for silicone rubbers is very low indeed, about $-125°$.

The properties of silicone and poly(propylene oxide) rubbers thus emphasize that there are *two* barriers to motion of chain segments in polymers. One is external and involves the attractive van der Waals forces between units in adjacent chains; at low temperatures it restricts intermolecular motion necessary to move polymer chains in a rubber. The other is internal and involves the restriction of rotation from repulsive van der Waals forces between adjacent units in the same chain; at low temperatures it restricts chain flexibility necessary to move polymer chains in a rubber.

SOLUBILITY OF POLYMERS

One of the factors favoring dissolution of any material is the accompanying gain in entropy. A pure solute and a pure solvent represent a more ordered molecular arrangement than a solution, in which the molecules are randomly mixed. As more and more atoms are joined together into large polymer molecules by bonds of fixed length and bond angles, the number of different independent arrangements these atoms can assume becomes more and more limited. Thus a dilute solution of 100 monomer molecules is a much more random array than a dilute solution of one polymer molecule containing 100 monomer units. In the former case, the 100 monomer molecules can wander through the solution independently. In the latter case, while the chain can assume many different conformations, the 100 monomer units must always remain near one another attached by covalent bonds.

Therefore, the entropy factor favoring solution decreases as the molecules increase in size, and polymers are, generally, appreciably less soluble than their separate monomer units. For covalent polymers, complete miscibility occurs in solvents having about the same CED parameter as the polymer, while increasingly limited solubility is found for solvents with much smaller or much larger CED. In general, the situation is governed by the two relations.

$$\Delta G_{sol} = \Delta H_{sol} - T \Delta S_{sol}$$

$$\Delta H_{sol} \propto (CED_{solute} - CED_{solvent})^2$$

This relationship predicts that the ΔH_{sol} will increase with the square of the difference in CED parameter, and this factor, being positive, is unfavorable to dissolution. The entropy term for solution ΔS_{sol} is also normally a positive number; i.e., the solution entropy is greater than that for the separate pure solute and solvent, and thus will favor solution, but less so for large than for small molecules of similar CED.

The entropy term is quite small if both solute and solvent are large molecules. Thus, the miscibility of one polymer with another is generally negligible unless, in fact, the CED parameters are nearly identical to permit such mixing. That is, since the entropy term will contribute very little favorably, there must be a very small unfavorable enthalpy term ΔH_{sol} if miscibility is to occur.

The considerations above must be modified considerably for the case of *crystalline* polymers. In a very general way, the chains in the liquid or rubbery state and in solution will tend to assume a randomly coiled conformation. The more the chain does assume a random coil, the higher its entropy will be. Thus chains which can easily assume random-coil conformations have a tendency to a lower melting point and a higher solubility.

One might compare polyethylene and polyisobutylene. They are both saturated hydrocarbons and thus have similar low cohesive forces between chains. Polyethylene, however, has a high crystal melting point and is resistant to virtually all solvents at room temperature. Polyisobutylene is amorphous when unstressed at room temperature and dissolves in a wide range of solvents with CED parameters in the neighborhood of 50 to 75 cal mole^{-1}. Both the low solubility and high melting point can be ascribed to the preferred trans conformation of the C—C bonds in polyethylene, leading to a relatively smaller entropy change for melting or dissolution

$$\Delta G_{sol} = \Delta H_{sol} - T \ \Delta S_{sol}$$

Since the ΔH_{sol} is not very different, a lower ΔS_{sol} means that dissolution (or melting) will be favored only at higher temperatures, to increase the $T \ \Delta S$ term. Therefore, it is understandable that polyethylene does indeed become soluble in a number of solvents of proper CED at temperatures above 100°C.

Another example of the relationship of high melting point to low solubility, other parameters being similar, is the case of isotactic and atactic poly(propylene oxide). The isotactic crystalline polymer (melting point $\sim 75°$) can be separated very neatly from the amorphous rubbery polymer of the same molecular weight by virtue of their great difference in solubility in cold acetone. At $-30°C$, the isotactic polymer is virtually insoluble while the amorphous polymer is highly soluble. Thus again, the structural features which promote crystallinity by decreasing the favorable entropy for melting decrease the favorable entropy for dissolution of the isotactic polymer compared to the atactic isomer.

This same marked difference in solubility is observed for other atactic and stereoregular polymers, such as atactic (amorphous, soluble) and isotactic (crystalline, insoluble) polypropylene.

In the discussions in this chapter, we have considered how the wide variety of physical properties of linear polymers, rubbery states, glassy states, fibers, widely differing melting points T_m, glass transition temperatures T_g, and solubilities, can be related to the nature of the units building the polymer chain. The symmetry of the units and their van der Waals interactions are the principal structural factors involved.

PROBLEMS

4.1 Recalling that the relation between freezing-point depression ΔT_f and molality m is $\Delta T_f = K_f m$, use the data in Table 4.1 to calculate K_f for the polyethylene crystal, using as m the molal concentration of polymer-chain end groups. Assume the density of polyethylene to be 1.0 (actually ~ 0.97).

4.2 Using the value of K_f for polyethylene crystals calculated in Prob. 4.1 and assuming a molecular weight of 10,000 and a melting point of 120°, calculate the number of chain branches per polyethylene molecule.

4.3 If $\Delta F_{str} = T \, \Delta S_{str}$ and $\Delta H_{fus} = T_m \, \Delta S_{fus}$, outline an experiment which could measure ΔS_{str} for different degrees of extension. What does the first equation say will happen to the tension on a stretched rubber band if the temperature is increased (at constant extension)? What other data would be needed to calculate the effect of degree of extension on T_m?

SUGGESTED READING

Billmeyer, F. W. Jr.: "Textbook of Polymer Science," chaps. 1, 5 – 7, Interscience Publishers, a division of John Wiley & Sons, Inc., New York, 1966.

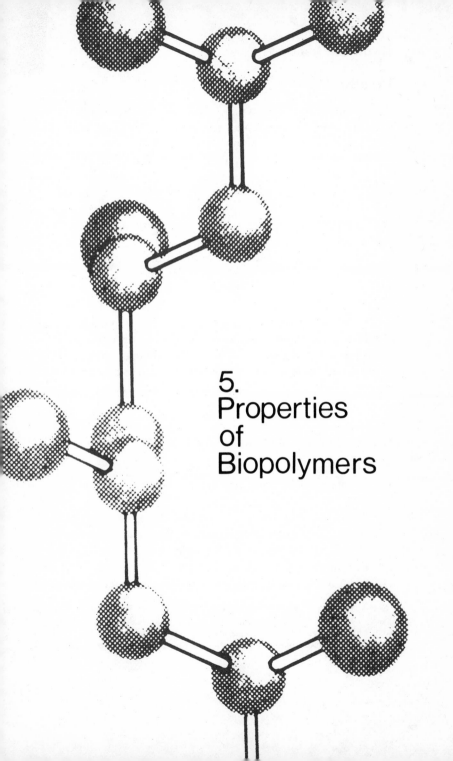

5.
Properties
of
Biopolymers

INTRODUCTION

The important biopolymers—nucleic acids, proteins, and carbohydrates—are essential for the functioning of every living cell. They are long-chain polymers and all are stereoregular; i.e., their units are chiral and have the same configuration, like isotactic polypropylene. These biopolymers are distinguished from those in the preceding chapter by the major and significant characteristic feature of very high CEDs, due to the high concentration of strong hydrogen bonds between units in these structures. The properties with which these polymers are endowed, thanks to these strong cohesive forces, are certainly important features of their biological functions. Furthermore, it is intriguing to speculate that the high cohesive forces of these polymer chains may have been a crucial feature in the origin of life, since such molecules, if formed in the primordial soup, would have a strong tendency to organization, so necessary in any concept of how the remarkably organized structure of a living cell, even a relatively simple primitive cell, could have derived from the disorganized dilute solution of organic material in the earliest oceans of earth.*

CELLULOSE AND STARCH

These two carbohydrates, major building material for plants and energy sources for animals, are both polymers of the sugar glucose.

Each 1,4 anhydroglucose unit in cellulose has the β configuration, i.e., the oxygen attached to C_1 is trans to the hydroxyl group at C_2. This also puts the linkage to the next unit at C_1 trans to the linkage to the next unit at C_4, as represented in the chair conformation in Fig. 5.1. Note that this puts all the groups

*See A. I. Oparin, "The Origins of Life," 2d ed., Dover Publications, Inc., New York, 1953; Sidney W. Fox (ed.), "The Origins of Prebiological Systems," Academic Press Inc., New York, 1965; D. Kenyon and G. Steinman, "Biochemical Predestination," McGraw-Hill Book Company, New York, 1969.

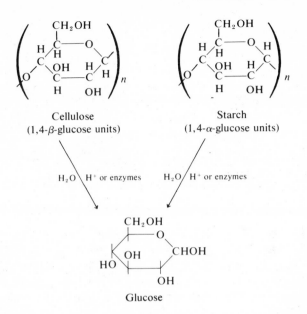

Cellulose
(1,4-β-glucose units)

Starch
(1,4-α-glucose units)

H_2O \ H^+ or enzymes

H_2O / H^+ or enzymes

Glucose

bulkier than hydrogen in the more stable *equatorial* positions, with all five hydrogens on the ring carbons in the *axial* positions. This also gives this unit an extended conformation and trans configuration which would fit with the linear rodlike nature of cellulose molecules in cellulose fibers, such as cotton or linen, and with the observed repeat distance of 10.25 Å along the fiber axis corresponding to the maximally extended chain.

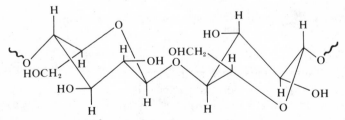

Fig. 5.1 *Adjacent 1,4-glucose units in cellulose fiber.*

These units not only have an extended geometry enabling them to form rodlike molecules preferred for fibers, but each unit has three hydroxyl groups, capable of forming hydrogen bonds to oxygens in other units.

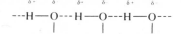

These bonds hold the molecules together with relatively very high van der Waals cohesive forces, and indeed the cellulose structure is so stable that it will not melt up to temperatures where the molecules themselves undergo thermal decomposition. It also is highly resistant to disruption by solvents.

Natural starch occurs in two modifications, which can be separated. On leaching starch grains at 70° with water, about 10 to 20 percent of the starch dissolves and then precipitates on cooling. This form is a linear structure called amylose; the remainder is a branched-chain structure called amylopectin. It involves not only the normal 1,4-α-glucose unit, but about every twenty-fifth unit there is also a glucose unit attached at the hydroxyl on C_6.

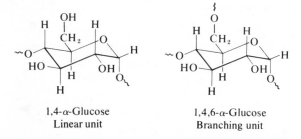

1,4-α-Glucose
Linear unit

1,4,6-α-Glucose
Branching unit

Amylose and amylopectin display only diffuse x-ray diffraction patterns indicating very little crystallinity. The chain geometry and the chain defects prevent the close association of interacting groups as in cellulose. However, the chain can take up in solution a quite unusual helical formation. The contrasting

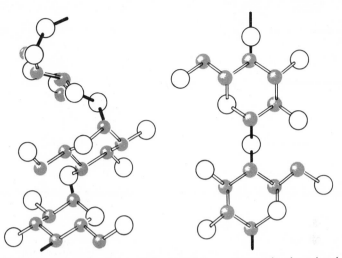

Fig. 5.2 *Starch and cellulose (left and right) are giant molecules related to α-D(+)-glucopyranose and β-D(+)-glucopyranose, respectively (hydrogen atoms omitted for clarity). Open circles = oxygen atoms; shaded circles = carbon atoms. (Redrawn from Giulio Natta, Precisely Constructed Polymers. Copyright © 1961 by Scientific American, Inc. All rights reserved.)*

arrangements in starch and cellulose are illustrated in Fig. 5.2 and the large helical structure of the starch chain in Fig. 5.3. This helix has about six glucose units for each turn, and the structure can be visualized as being held together by strong hydrogen bonds between the free hydroxyl groups in a glucose unit and those in the units in the turn of the chains above or below it. The dimensions of the starch helix leave a hole down the center of the helix, which however, normally holds water molecules, one for each glucose unit in the chain. The hole turns out to have just the proper dimensions to accommodate an iodine molecule, I_2, about one molecule per turn of the helix. The iodine molecules in this environment assume a deep blue color, which is a sensitive and characteristic indicator for the presence of either iodine or starch.

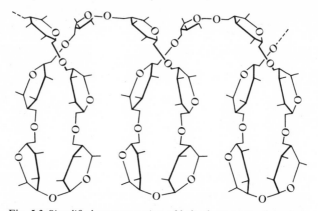

Fig. 5.3 *Simplified representation of helical structure of amylose in solution. The molecular shape of amylose explains some of its chemical and biochemical properties. With iodine, amylose turns intensely blue. This venerable reaction has been explained only recently, when it was found that iodine molecules enter the hollow center of the coil created by the glucose units. In such enclosure compounds iodine exhibits changed physical characteristics, such as the strong absorption of light. There are enzymes that slit open this tube and free fragments, each of which contains six glucose units in a ring. (Redrawn from P. Karlson, "Introduction to Modern Biochemistry," p. 305, Academic Press, Inc., New York, 1965.)*

PROTEINS

These polymers are major building material for animal structure, such as skin, tendon, and muscle, and are also the essential ingredients of that remarkable array of catalysts present in every living cell and responsible for the amazingly specific and efficient chemical reactions necessary to the life processes. Many different proteins have been isolated in pure crystalline form; a limited number have now had their complete complex crystal structure deciphered by x-ray methods; and a few, such as insulin and ribonuclease, have been synthesized in the laboratory.

The major building blocks of the proteins are 20 α-amino acids; the structures are represented in Table 5.1.

Table 5.1 Structures of 20 Amino Acids of Proteins

R	Name
1. $H-$	Glycine
2. CH_3-	Alanine
3. $(CH_3)_2CH-$	Valine
4. $(CH_3)_2CHCH_2-$	Leucine
5. $CH_3CH_2\underset{\underset{CH_3}{\mid}}{CH}-$	Isoleucine
6. $C_6H_5CH_2-$	Phenylalanine
7. $HOCH_2-$	Serine
8. $CH_3\underset{\underset{OH}{\mid}}{CH}-$	Threonine
9. $CH_3SCH_2CH_2-$	Methionine
10. $HSCH_2-$	Cysteine
11. $\underset{HO}{\overset{O}{\diagdown}}CCH_2-$	Aspartic acid
12. $\underset{HO}{\overset{O}{\diagdown}}CCH_2CH_2-$	Glutamic acid
13. $H_2NCH_2CH_2CH_2CH_2-$	Lysine
14. $H_2NCH_2\underset{\underset{OH}{\mid}}{CH}CH_2CH_2-$	Hydroxylysine
15. $\underset{H_2N}{\overset{HN}{\diagdown}}CNHCH_2CH_2CH_2-$	Arginine
16. $\underset{HC=C-CH_2-}{\overset{H}{\overset{C}{N\diagup\diagdown NH}}}$	Histidine

*Except for 19, proline, and 20, hydroxyproline

Table 5.1 *(Continued)*

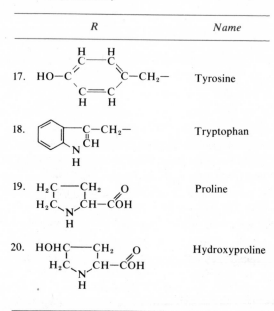

R	Name
17.	Tyrosine
18.	Tryptophan
19.	Proline
20.	Hydroxyproline

Except for proline and hydroxyproline (19 and 20 in Table 5.1), these amino acids can be represented by the general formula

$$RCH \underset{\diagdown COOH}{\overset{\diagup NH_2}{}}$$

and the polymers as

$$\left(NHCHCO \right)_n \overset{R}{}$$

Proteins can be hydrolyzed to their component amino acids, either enzymatically (as in digestion) or by hot acids.

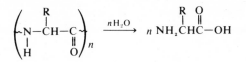

Amino acid unit
in a protein

α-Amino acid

Note that all 20 amino acids except glycine have an asymmetric α carbon atom and are thus optically active. All those normally isolated from proteins have been shown to have the same configuration at the α carbon atom. In this geometric or stereochemical sense, the protein chain can thus be considered to be analogous to isotactic polypropylene.

Like isotactic polypropylene, one of the preferred conformations of the protein chain is a helix, as first proposed by Linus Pauling for the α helix (see Fig. 5.4). The helical arrangement is stabilized in this case by the effective hydrogen bonding between the hydrogen of the NH groups and the oxygen of the C=O groups. The tendency to an orderly arrangement is facilitated by the following resonance interaction, which contributes to the structure of the amide group and which, due to the

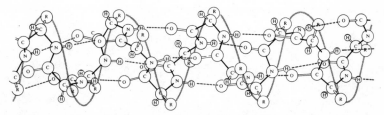

Fig. 5.4 *An α helix gives a polypeptide chain the linear structure shown here in three-dimensional perspective; —, bonds in peptide chain; =, bonds to peptide chain. The atoms in the repeating unit (CCONHC) lie in a plane; the change in angle between one unit and the next occurs at the carbon to which the side group is attached. (Redrawn from Paul Doty, Proteins. Copyright © 1957 by Scientific American, Inc. All rights reserved.)*

double-bond character of the N—C bond, imposes a preferred planar conformation for the six atoms indicated below.

This planar arrangement at the peptide bond

means that the helical conformation of a protein chain is due to skew conformations for the two chain bonds at the α carbon

The contribution of the second more polar electron arrangement also enhances the negative charge on the oxygen and the positive charge on hydrogen, thus strengthening the hydrogen bonding which binds coils in the helix together (dashed red lines in Fig. 5.4).

Another important arrangement for protein molecules is the

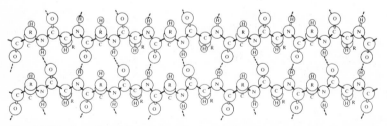

Fig. 5.5 *The β configuration ties two or more polypeptide chains to one another in crystalline structures. Here the hydrogen bonds do not contribute to the internal organization of the chain, as in the α helix, but link the hydrogen atoms of one chain to an oxygen of the next. (Redrawn from Paul Doty, Proteins. Copyright © 1957 by Scientific American, Inc. All rights reserved.)*

β configuration, shown in Fig. 5.5. In this arrangement, the polymer chains are extended, and hydrogen bonding occurs between amide linkages in adjacent chains rather than adjacent units in the helical turns of the same chain, as in the α helix.

The actual crystal structure of many natural proteins has been shown to be far more complex than a simple long α helix or extended chains in the β form. The structures for hemoglobin, lysozome, insulin, and ribonuclease (see Fig. 5.6*) reveal they are *globular*, rather than linear or helical in overall shape. At least one reason for this is the S—S cross-linking which occurs between cystine units in the sequence, as in the insulin molecule. These —S—S— cross-links in essence "vulcanize" the protein chain into somewhat preferred conformations, just as —S— and —S—S— cross-links do in vulcanized rubber.

Because of the complexity of proteins, their structural organization is described at different levels, or degrees, known as primary, secondary, etc., structure. The most fundamental, and the first to be clearly established for any pure protein, is the precise sequence of the amino acids of which it is composed. The pioneer example was the establishment of this *primary* structure of insulin by Sanger in 1952.

In addition to the primary peptide sequence, for any protein containing cystine units, it is necessary to establish exactly which cysteine units have been cross-linked by a S—S bond.

$$\underset{\text{Cysteine}}{2\text{HSCH}_2\overset{\overset{\displaystyle \text{NH}_2}{|}}{\text{CHCOOH}}} \underset{\text{redn}}{\overset{\text{oxid}}{\rightleftharpoons}} \underset{\text{Cystine}}{\begin{array}{c}\overset{\displaystyle \text{NH}_2}{|}\\ \text{SCH}_2\text{CHCOOH}\\ |\\ \text{SCH}_2\text{CHCOOH}\\ |\\ \text{NH}_2\end{array}}$$

*The reader is referred to "Bio-organic Chemistry," readings from the *Scientific American*, W. H. Freeman and Company, San Francisco, Calif., 1968, pp. 64 and 65 for a three-dimensional representation of lysozome and pp. 24 and 25 for the schematic structure of ribonuclease, and to The Synthesis of Living Systems, *Chem. Eng. News*, **45**:144–156 (1967), for the structure of insulin.

Fig. 5.6 *Hemoglobin molecule, as deduced from x-ray diffraction studies, is shown from above (p. 106) and side (p. 107). The drawings follow the representation scheme used in three-dimensional models built by Perutz. The irregular blocks represent electron-density patterns at various levels in the hemoglobin molecule. The molecule is built up from four subunits; two identical α chains (light blocks) and two identical β chains (dark blocks). The letter N in the top*

These sulfur cross-links impose a major restriction on the possible conformations of the protein chain and can, for example, clearly block it from assuming a regular helical-rod conformation. Moreover, other specific noncovalent interactions among some amino acids can have the same effect. The third level of protein organization is determined by the extent to which van der Waals forces hold the chains together in helical segments or other conformations. For many globular proteins, there is then a

view identifies the amino ends of the two α chains; the letter C identifies the car-boxyl ends. Each chain enfolds a heme group (colored disk), the iron-containing structure that binds oxygen to the molecule. (From M. F. Perutz, The Hemoglo-bin Molecule. Copyright © 1964 by Scientific American, Inc. All rights reserved.)

fourth level of organization, the detailed packing of a few protein molecules to form specific groups such as tetramers. These then become the units from which the crystal is built.

One critical question about the complex levels of protein chain organization has been recently answered by the syntheses of the important proteins insulin and ribonuclease. One could wonder whether the remarkable specific properties of these sub-stances—insulin to regulate carbohydrate metabolism and ribonuclease to catalyze the rapid hydrolytic destruction of RNA—were dependent not only on the precise proper sequence of amino acids but a unique chain conformation. If the latter were imposed by some special process of the living cell, it might be possible for a chemist to reproduce the proper sequence but still not have duplicated the properties of insulin or ribonuclease. It turns out that the synthetic materials in each instance not only are identical with the natural ones in gross chemical and physical properties but show identical capacity to regulate carbohydrate metabolism and RNA hydrolysis, respectively. There is thus no magical "vital force" which shapes these protein chains; rather, their conformation is one naturally preferred by the chain due to its primary sequence of amino acids, its resulting exact location of cystine cross-links, and the areas of bulky and/or hydrophobic or hydrophilic groups in the groups R attached to the chain backbone. For example, side chains with polar groups like aspartic acid, glutamic acid, arginine, and serine would contribute relatively high CEDs, and segments rich in these units would enhance water solubility [hydrophilic (water-loving)]. Large hydrocarbon side chains such as leucine, isoleucine, and phenylalanine would have relatively low CEDs, and segments rich in these units would resist dissolution in water [hydrophobic (water-hating) or lipophilic (fat-loving)]. Sequences which lead to accumulations of either hydrophilic or lipophilic groups in proteins will influence the preferred chain conformation and are believed to be a vital factor in determining the remarkably specific catalytic activity of many enzymes.

NUCLEIC ACIDS

These polymers are formed from a more complex unit than the other biopolymers, involving phosphoric acid, a sugar [ribose for RNA (*ribo*nucleic *a*cid) and deoxyribose for DNA (*deox*yribo*n*ucleic *a*cid)], and a complex organic base such as adenine, guanine, cytosine, thymine, or uracil. One of the triumphs of the last few decades has been the deciphering of the general structure of the nucleic acids and the role they play in the transcription of genetic information. The sequences in DNA molecules dictate those for messenger and transfer RNA, which then program the sequences of all the proteins necessary for the functioning of a living cell as the amazing chemical factory it really is.

The building blocks of nucleic acids are shown in Fig. 5.7. The bases in DNA are adenine, thymine, guanine, and cytosine. In RNA, thymine is replaced by uracil, identical to thymine except for a hydrogen in place of the methyl group, CH_3, of thymine. One of the key features of the composition of various samples of DNA is that adenine and thymine are usually present in equimolar amounts, and guanine and cytosine also. For example, if there is 28 mole percent of adenine, there will be 28 mole percent of thymine and 22 mole percent of guanine and 22 mole percent of cytosine. One of the virtues of the double-helix structure for DNA, discovered by Watson and Crick in 1953, was its explanation of this relationship, inasmuch as the two

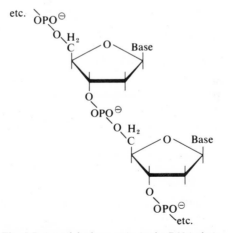

Fig. 5.7 *A model of two units in the DNA chain. (From T. L. V. Ulbricht, "Purines, Pyrimidines and Nucleotides and the Chemistry of Nucleic Acids," p. 65, The Macmillan Company, New York, 1964.)*

chains are held together by base pairing through hydrogen bonds between adenine and thymine units and between guanine and cytosine units, as illustrated in Fig. 5.8.

The stereoregular configuration at each sugar ring also contributes to a preference for a regular helical conformation. One remarkable feature of the double helix of DNA is that it persists unchanged up to relatively high temperatures (80° in aqueous sodium chloride solutions). At a temperature nearly as sharp as a melting point for a microcrystal this highly organized double-helix structure will "melt," and the DNA molecules become separate random coils. The temperature for this transition from the *natural* double helix to *denatured* random coils depends on the solvent. For example, it changes depending on the salt concentration and pH. The thermodynamics of the pro-

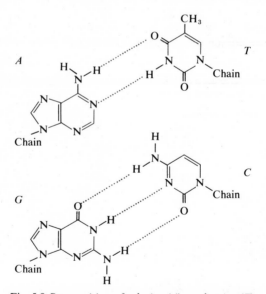

Fig. 5.8 *Base pairing of adenine (A) to thymine (T) and guanine (G) to cytosine (C) units in DNA. (From T. L. V. Ulbricht, "Purines, Pyrimidines and Nucleotides and the Chemistry of Nucleic Acids," p. 66, The Macmillan Company, New York, 1964.)*

cess is similar to melting, so that the transition temperature T_m must be the temperature defined by the enthalpy and entropy of denaturation by

$$\Delta H_{\text{den}} = T_m\, \Delta S_{\text{den}}$$

In the more detailed model of a segment of the DNA double helix in Fig. 5.9, note that the planar base molecules fit together in the core of the helix with their faces parallel to each other and at right angles to the axis of the helix. The strong van der Waals association engendered by this favorable orientation of the base units is undoubtedly as important as the hydrogen bonding (Fig. 5.8) in holding the double-helix structure together.

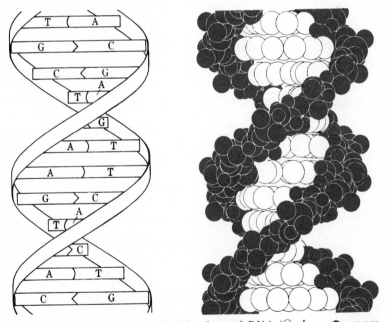

Fig. 5.9 *A model of the double-helix form of DNA* ($\bigcirc$, *base;* $\bullet$, *sugar-phosphate chain). (From Sienko and Plane, "Chemistry," 4th ed. Copyright* © *1971 by McGraw-Hill, Inc. Used with permission of McGraw-Hill Book Company.)*

The process of denaturation can be measured readily by following the light absorbance in the ultraviolet at 260 nm since the absorbance by denatured DNA is 40 percent greater than for the natural double helix. When microbial DNA is denatured by heating and the solution cooled, considerable *renaturation* occurs; i.e., many of the chains reassemble into the double-helix arrangement. When animal DNA is similarly denatured and cooled, virtually *no* renaturation is observed. One explanation for this difference is the far more complex genetic information in animals, so that every DNA chain (of thousands in every cell) may contain quite different base sequences, whereas the relatively simpler genetic information in microbial DNA means that many of the various DNA molecules are similar or even identical. Since to reform a double helix, proper base pairing demands precisely matched chains, once the animal DNA chains melt from the double-helix units to single random-coiled chains, it is extremely improbable that pairs will find the proper partner among the thousands of chains with differing sequences.

However, it has been shown that renaturation of animal DNA is possible if the chains are prevented from becoming entirely free to separate from each other after denaturation. Rutman has shown that calf thymus DNA, cross-linked between guanine units by nitrogen mustard, denatures at exactly the same temperature as the untreated DNA, but whereas the untreated DNA fails to renature on cooling, the cross-linked DNA does renature. Furthermore the extent of renaturation increases with the amount of nitrogen mustard used to cross-link the chains.

Two guanine units in
DNA chains

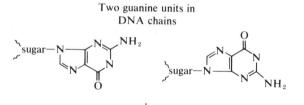

+

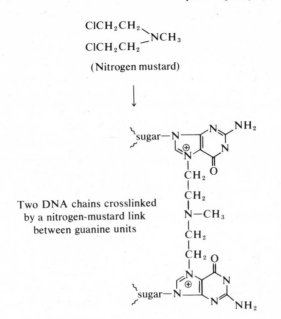

$ClCH_2CH_2$
$ClCH_2CH_2$ NCH_3

(Nitrogen mustard)

Two DNA chains crosslinked
by a nitrogen-mustard link
between guanine units

It is our hope that this volume will have introduced the student to material relating the properties and structures of simple molecules to those of the important natural and synthetic long-chain polymers. The behavior of biopolymers will be further developed in another volume in this series by Paul M. Doty.

SUGGESTED READING

"Bio-organic Chemistry," readings from the *Scientific American*, sec. I, W. H. Freeman and Company, San Francisco, Calif., 1968.

Index